Causality and Physical Theories
(Wayne State University, 1973)

AIP Conference Proceedings

Series Editor: Hugh C. Wolfe

Number 16

Causality and Physical Theories
(Wayne State University, 1973)

Editor

William B. Rolnick

Wayne State University

American Institute of Physics

New York   1974

L.C. Catalog Card No. 73-93420
ISBN 0-88318-115-0
AEC CONF-73056

American Institute of Physics
335 East 45th Street
New York, N.Y. 10017

Printed in the United States of America

EDITOR'S FOREWORD

This conference was held at Wayne State University on May 11-12, 1973. It was funded by the Wayne State University Alumni Association which provided a Research Recognition Award. I wish to thank Dr. Cyrus Moazed who helped me organize and run the conference, Mr. K.C. Lee who did the drawings, and the Physics Department secretaries for the typing. I thank my wife Ellen for her patience and encouragement.

TABLE OF CONTENTS

Introductory Remarks on Causality
    William B. Rolnick.. 1

Historic Views of Causality
    Richard Schlegel... 3

Causality and Relativistic Dynamics
    Peter Havas... 23

Relativity and the Order of Cause and Effect in Time
    Roger G. Newton... 49

General Physical Principles and Non-Linear Group Relations
    Max Dresden... 65

Macrocausality and Its Role in Physical Theories
    Henry P. Stapp.. 87

Quantum Theory With Shadow States: A Separate Reality
    E.C.G. Sudarshan....................................... 115

Causality and Metrical Properties of Matter in a Two-Metric
  Field Theory of Gravity
    K. Nordtvedt, Jr. 137

Tachyons, Causality, and Rotational Invariance
    Allen E. Everett and Adel F. Antippa................... 147

Indeterminism, Time Arrow, and Prediction
    F.J. Belinfante....................................... 157

List of Participants....................................... 176

INTRODUCTORY REMARKS ON CAUSALITY

William B. Rolnick
Wayne State University, Detroit, Michigan 48202

The concept of causality has caught the imagination of man over the centuries, and has inspired much thought and controversy.

The purpose of this conference is to explore causality and its interaction with our theories of the physical world. Our concept of causality is very much intertwined with our philosophical feelings about the world, with our basic feelings about reality and experiments. Although causality is an elusive concept, we each have an intuitive feeling about it.

In past years, the traditional concept of causality has changed as a result of relativity and quantum theory, and more recently a number of theories have been created which seem to be altering our concept of causality even more. Tachyon theories are among those, and I mention them to illustrate a serious misconception which has occurred. There are tachyon theories which work very well with no contradictions if you have a closed four dimensional space-time continuum. We can construct a theory with advanced potentials, or advanced Green's functions, and there will be no contradictions. There have appeared in the literature many such examples, but in those cases causality doesn't exist because when you have a closed four dimensional space-time manifold, you merely have events laid out on this manifold and there is no identification of one event being the cause of another. It is only when you have an open world in the sense that you can envision yourself doing an experiment and creating a disturbance that causality becomes a meaningful concept. The disturbance will represent a cause to you and then you look for the effect of that disturbance. Theories with Green's functions which contain advanced parts or tachyonic modes seem to lead to problems with causality (e.g., logical contradictions) when we allow "experiments" to be done by an agent outside the system. (In an open world, one can get around certain objections to tachyon theories by introducing preferred reference frames, etc.)

In this conference, we will attempt to answer, or at least refine, the following questions: First of all what do we mean by causality and how do the presently proposed physical theories change our basic concept of causality. Secondly, should our philosophical idea of causality present any restrictions on our theories, or, is causality just a luxury so that, as we develop more and more theories and do more and more experiments, we may change it as much as is necessary. Thus, the main purpose of this conference is to look at the theories which seem to have proposed changes in our ideas of causality and also to investigate whether there are certain basic ideas of causality which are unchangable and therefore will restrict our theories.

HISTORIC VIEWS OF CAUSALITY

Richard Schlegel
Department of Physics, Michigan State University
East Lansing, Michigan 48823

ABSTRACT

Causality has a central role in physical thought;
but there is wide variation, historically, in what is as-
cribed to particular causal properties such as degree of
determinism, past-future asymmetry, and requirement of
physical plausibility. Doctrines of causality in
Aristotle, Newtonian mechanics, David Hume, Bertrand
Russell, and current quantum theory are examined. Although philosophical analysis is a determinant of views
of causality, it is apparent that there is also a strong
interdependence between what has been established in
science and what is accepted as the principle of causality.

INTRODUCTION

Accepted assertions in any given historical period about what
constitutes cause-effect relations must to a considerable degree reflect ideas about what nature is and how it operates. A survey of
past doctrines relating to causality should, therefore, tell us
something of past developments in physics, as well as give indica-
tions of how we have been brought to the notions of causality that
we now accept.

Our treatment could, obviously, go into many details of people
and ideas, in view of the wealth of available historical material.
My procedure for attaining a discussion of appropriate length will
be to give close attention to only three major writers on causality:
Aristotle, David Hume, and Bertrand Russell. In a final section
the influence of quantum theory on ideas of causality will be discussed.

I. ARISTOTLE

In the second book of his Physics[1] - which is essentially an
exhaustive study of motion, including also such topics as time,
place, the infinite - Aristotle presents his celebrated statement
of the four kinds of cause. He first points out[2] that our inquiry
is for the sake of understanding, and we do not understand a thing
until we have acquired the why of it, the proximate cause; this we
must do, then, with respect to generation and destruction and every
physical change. He finds that there are four different senses of
cause:

1. Material Cause. This is that which is a constituent of
   that which is generated. Thus, Aristotle writes that sil-
   ver may be the material cause of a cup and bronze is a
   cause of the statue. We today could say that protons,
   neutrons, and electrons are the material cause of the
   atom; and that we do not know the material cause of the
   electromagnetic field.

2. <u>Formal</u> <u>Cause</u>. In another sense, cause is the form or pattern, or the structure ("formula of the essence"). Thus, the formal cause of a house is the design. Aristotle gives the example, the formal cause of the octave is the ratio 2:1. I presume we could say that the Schrödinger equation, specified for the electron and proton in electrostatic interaction, is the formal cause of the H atom.

3. <u>Efficient</u> <u>Cause</u>. In a third sense, cause is that which initiates change or motion or coming to rest. A moving bat is the efficient cause of a ball's motion, as is an electric field for a current. Aristotle says the father is the efficient cause of the baby (I suppose we could add that for the most part the comparable role of the mother is to be the material cause, and that the genetic structure in egg and sperm cell is the formal cause).

4. <u>Final</u> <u>Cause</u>. This is that for the sake of which the change occurs. We walk for the sake of health, Aristotle says, and so health is the final cause of the walking. The use to which the house is put is the final cause of building it. To be a physician might be the final cause for a student's being in college.

Aristotle points out that causes are spoken of in many ways, and the same cause may be both prior (cause) and posterior (effect); thus, 2 is a cause for octave, and "number" may be a cause for 2, as a form of all numbers. Also, a pair of causes may each be cause of the other; for example, exercise is a cause (efficient) of good physical condition, whereas good physical condition is a cause (final) of exercise.

Even with the passage of some twenty-three hundred years Aristotle's doctrine of the four causes remains as the single greatest contribution to our understanding of causality. It is, I believe, still perceptive and useful in telling us what we do in science, and if I were writing a textbook in physics I would include a statement of the doctrine. It is true that in physics we usually are not concerned with final causes; that may be a defect, and in any event it would be well to be aware of the lacuna. Certainly, biology and the social sciences are to some degree concerned with final causes.

II. TRANSITION TO A UNIVERSAL MECHANICS

We pass over many centuries now, to the development of the Newtonian mechanics that gave a powerful ideal and model for causation in all investigations of the natural world. But first we say a few words about the transition.

For Aristotle there were two causes for motion of an object: either something external or a principle of motion within it. Thus, a stone moves on the ground because it is pushed, but it falls to the earth because there is an inner principle of motion that causes all terrestrial bodies to move toward the center of the earth. (Likewise, there is a principle that causes astronomical bodies to move in circles.) Obviously, there is some wrong physics here,

and questions that are difficult to answer arise with the Aristotelian
theory. For example, how is the horizontal motion of an arrow to
be explained when the bow string is no longer forcing it? The
answer given by the Aristotelians was that arrows and projectiles
continue their motion because air that is pushed and compressed in
front moves around to the rear, filling the vacuum that would otherwise be left by the moving object and also exerting a forward push.
Herbert Butterfield[3] points out that Renaissance artillerymen were
still victims of this error, even though late medieval philosophers
had come to a better theory.

The improved theory, which can be said to constitute the beginning of Newtonian mechanics, came at the University of Paris in
the 14th century, notably with Jean Buridan, Albert of Saxony, and
Nicholas of Oresme.[4] In their discussions of motion, the Parisian
school came to the doctrine of _impetus_, to the transfer of the principle of motion from an external force to one of inertia or momentum that is within the body itself. Impetus stays in the body, as
does heat in a poker after it is taken from the fire, and gives continued motion; for falling bodies there is acceleration because
gravity adds impetus. Thus, a long first step was taken toward
Newton's laws of motion.

In the 17th and 18th centuries the system of what we now call
classical mechanics was established; not only with a mathematical
formulation that - in the minds of many - could eventually lead to
a description of all the natural world and its processes, but with
a central role for physical force. This could be directly and in-
tuitively identified with the immediately known kinesthetic force
of human experience, thereby giving a seeming confirmation of the
universal mechanics.

The deterministic and apparently completely adequate natural
philosophy that many accepted as demonstrated by classical mechanics was formulated by Pierre Simon de Laplace: "We ought then to
regard the present state of the universe as the effect of its an-
terior state and the cause of the one to follow ... an intelligence
which could comprehend all the forces by which nature is animated
and the respective situation of the beings who compose it ...
would embrace in the same formula the movements of the greatest
bodies of the universe and those of the lightest atom; for it,
nothing would be uncertain and the future, as the past, would be
present to its eyes. The human mind offers, in the perfection
which it has been able to give to astronomy, a feeble idea of this
intelligence."[5]

III. HUME AND CAUSATION

A. In accordance with the view of the world-as-mechanism, two
leading British philosophers, John Locke (1632-1704) and George (Bishop)
Berkeley (1685-1753), took the view that causality is a relation making
for motion or change. "Quite in conformity with the spirit of the age,
both of them admitted only the efficient cause of Aristotle. The other
three kinds of causes were considered 'too metaphysical'. Causality
is efficacy or efficient causality and this belongs to substances".[6]

So we find with the Newtonian world view a severe limiting of the causality concept: things causing other things to move becomes the mode of understanding the natural world. This point of view became the dominant one in natural science, and is perhaps still so today for the man in the street of the industrialized world. Already in the 18th century, however, the idea of efficient cause as a firm basis for philosophy of nature received a searching criticism from David Hume (1711-1776). After Aristotle, his ideas must be regarded as having the greatest influence on the philosophical views of causality.

B. Hume divides[7] objects of human reason into two kinds: relations of ideas and matters of fact. The former gives us certain truths as in geometry, algebra, and arithmetic; the latter (matters of fact) arise from experience, and, rather than being undoubted certainties, are such that we can conceive their opposite as being intelligible. Thus, we cannot conceive in Euclidean geometry of the sum of the angles of a triangle as being other than 180°, but we can readily conceive of circumstances under which the Sun would not rise tomorrow. Philosophers express some reservations, but we still live today substantially with this same distinction, between what are often called analytic and synthetic propositions.

All reasoning concerning matters of fact seems to be founded, Hume says, on the relation of Cause and Effect; for, it is with that relation that we can go beyond the evidence of our senses. Thus, you believe that a person has recently been in a room because you find a fire in the fireplace. All causes and effects are discovered, however, by experience; who can imagine, say, that the attraction of iron for loadstone could be established other than by experience.

There is, however, Hume writes, some disposition to think that it is otherwise with events which have become very familiar to us. We are apt to imagine that we could, e.g., have inferred that one billiard ball would on impulse communicate motion to another. But reflection will convince us otherwise. For, the effect is totally different from the cause, and can never be discovered within it. Motion in the second billiard ball is quite distinct from motion in the first, and there is nothing in one to suggest the smallest hint of the other. A stone left unsupported in the air falls, but is there anything we discover in this situation which can beget the idea of a downward rather than upward motion? When a given object (event) has always been found to be attended with a certain effect one can foresee that similar objects will be attended with similar effects. But such an inference is not based on any chain of reasoning, and Hume challenges anyone to produce any argument which supports by reason the drawn inference.

C. People argue that there is necessity, or secret powers, or efficacy, or agency, or connection, or productive quality, which require that the effect follow the cause. The key point of Hume's argument is that no such entities are observed or justifiably inferred. All we observe is Cause-Effect sequence, and we have no reasonable grounds for seeing it as necessarily obtaining at a future observation. He writes:[8] "There is, then, nothing new either discover'd or produc'd in any objects by their constant conjunction, and by the uninterrupted resemblance [repetition] of their relations of succession and contiguity. But 'tis from this resemb-

lance, that the ideas of necessity, of power, and of efficacy, are
derived. These ideas, therefore, represent not anything, that does
or can belong to the objects, which are constantly conjoined ....
The idea of necessity arises from some impression. There is no impression convey'd by our senses, which can give rise to that idea....
It must, therefore, be deriv'd from some internal reflexion....
Upon the whole, necessity is something that exists in the mind, not
in objects, ... necessity is nothing but that determination of the
thought to pass from causes to effects and from effects to causes,
according to their experienced union."

Later in this same section Hume writes, "I am, indeed, ready
to allow, that there may be several quantities both in material and
inanimate objects, with which we are utterly unacquainted; and if
we please to call these power or efficacy 'twill be of little consequence to the world. But when, instead of meaning these unknown
qualities, we make the terms of power and efficacy signify something, of which we have a clear idea, and which is incompatible
with those objects, to which we apply it, obscurity and errors be-
gin then to take place, and we are led astray by a false philosophy.
This is the case, when we transfer the determination of the thought
to external objects, and suppose any real intelligible connection
between them; that being a quality, which can only belong to the
mind that considers them."

D. So, Hume is telling us that there is no necessity of
causal connection in nature - only repeated sequences of events to
which we apply the name cause-and-effect. Hume removes, then, a
generally assumed underpinning from the philosophical view that
went with the mechanical philosophy: that there are in nature inherent laws or restrictions which require the behavior of mechanical systems. I like to imagine a scene in which Hume talks to a
natural philosopher of his day who demonstrates that balls, impelled
upward by a spring compressed to a fixed length, will rise only as
high as the energy given by the spring will propel them: only to
that height, we would say, at which the gravitational potential
energy is equal to the initial kinetic energy. The experimenter
might have to accede in logic to Hume's argument that no necessity,
only past experiences, are in support of the height limitation.
But at heart he might well feel, "I know that there is a necessary
element in nature which keeps the ball from rising higher". And
yet, Hume's argument now seems to have the last word. For, we know
from quantum physics that a mass particle in fact can go into
spatial regions where its   energy is less than the potential
energy associated with its presence in the region.

We see in David Hume's work a far removal of cause-effect from
intuitive notions associated with volition ("I cause the table to
move") and kinesthetic impression ("I feel the force that moves the
table"). For Hume, cause-effect becomes a succession of certain
states in time, and nothing more as far as the objects in the
natural world are concerned. Cause-effect are also, for Hume,
ideas in the mind, and he places their necessary connection there.
We can go two ways now: one would be to see how the man who is

generally considered the greatest philosopher of the modern period, Immanuel Kant, developed the notion of causality as having its source and being in the mind. This excursion into German idealism would indeed be of interest, particularly in view of the significant element of subjectivism that has come into physics with the quantum theory. Such an investigation does ask to be made. But the story would have to be long and my own superficial, tentative conclusion is that the fit is not a good one: the superposition of microscopic states which leads to observer-dependence in quantum physics seems to be different in kind from the a priori role of ideas in the Kantian philosophy. We shall, therefore, take the other route and continue the discussion of a generally empirical, scientific view of causality, as given by another British philosopher, of a stature comparable to that of Hume.

IV. BERTRAND RUSSELL

A. With typical iconoclasm, Russell asserts[9] of the word "cause" that it is so bound up with misleading associations that its complete extrusion from the philosophical vocabulary would be desirable. The term is never found in advanced sciences, such as celestial mechanics, and philosophers should bow to physicists on the matter and accept that the law of causality "is a relic of a bygone age, surviving, like the monarchy, only because it is erroneously supposed to do no harm".[10]

Russell does not deny that regularities occur: thus, stones break glass, if the momentum of the striking stone is great enough. Observed regularities of such a kind may be useful to a science in its infancy, but he denies that science assumes the existence of such invariable uniformities between observed events or that it aims at discovering them. In fact, he writes, "same cause, same effect" is of no importance to scientists; when all antecedents are known sufficiently well to enable a calculation to be made, they are so complicated that it is unlikely that they will ever recur. With this principle science would remain sterile. Also, causes do not "operate". Russell, in a Humean way, attacks this notion as being but an analogy with human volition (which does operate).

B. The supposed law of causality is replaced by the following principles:[11]
   1. In the case of frequently observed sequences we may say that the earlier event is the cause of the later event. But the sequence is only probable, not necessary as in the law of causation.
   2. It will not be assumed that every event has some antecedent which is its cause in the sense of 1. We shall only believe in causal sequences where we find them, without any presumption that they are always to be found.

3. <u>Any</u> case of sufficiently frequent sequence will be causal in our present sense; e.g., we shall not refuse to say that night is the cause of day.

4. Laws of probable sequence, although important in daily life and in the early stages of a science, tend to be displaced by quite different laws as science advances. Thus, in gravitational theory there is no cause and no effect: in the theory of motions of mutually gravitating bodies there is only an equation. The law of cause and effect is replaced by a <u>functional dependence</u>; and the constancy of scientific law consists in the <u>sameness of equations</u>, not in the sameness of causes and effects. If the "law of causality" is to be something actually discoverable, this should be its content.

The concept of functional dependence is the key element in Bertrand Russell's treatment of causation. He makes, among others, the following comments on the concept.[12] First, functional dependence is not <u>a priori</u> or self-evident or a "necessity of science"; it is not at all a premise of science, but is an empirical generalization from a number of laws which are themselves empirical generalizations. Second, the principle of functional dependence makes no distinction between past and future. The "future" determines the "past" to the same degree that the "past" determines the "future". And, third, something like the "uniformity of nature" (rather than the old "law of causality") is accepted on inductive grounds. This principle does not assert "same cause, same effect", but, rather, the permanence of laws of nature. The ground of the principle is simply "that it has been found to be true in very many instances; hence the principle cannot be considered certain, but only probable to a degree which cannot be accurately estimated."

C. Russell's view of causality as functional dependence has been highly influential among philosophers of science (although by no means everywhere uncritically accepted). We can see that it is, obviously, a further step in removing notions of specific mechanism or "reason why" from natural processes; the functional dependence is abstract, mathematical, and does not require intuitive understanding of why the relation obtains. Further, the idea that "cause precedes effect in time", which seems so basic to traditional causality, is lost.

It could be said that Russell's view is philosophically sophisticated, and even in advance of scientific thinking. Thus, his relaxing of the before-after requirement for cause-effect has appeared explicitly in scientific thinking, as in the Wheeler-Feynman action-at-a-distance electrodynamics.[13] In that theory, the radiative reaction of an accelerated charged particle is calculated with use of the assumption that an advanced potential has been emitted so as to arrive at the particle just at the time of acceleration.[14] One may well wish to argue against such an indifference of causal sequence to past-future relations. Thus, I would suggest that there is a time order determined by the processes of the natural world, and to put causes ahead of effects is to violate that order.[15] However,

the point obviously is moot, and if we are not committed to a temporal pattern of causal influence, as Russell says we should not be,
a theory such as that of Wheeler and Feynman is unobjectionable on
causality grounds.

   D. I wish also to argue however, that there is a valid content
to "causal relation" that goes beyond mere proximate concomitance of
appearance in a functional dependence. To say, as Russell approv-
ingly does, that night is the cause of day, because day invariably
in our experience follows night, is to admit too loose a definition
of cause and effect. We do know why night and day alternate: the
rotation of the earth and its opaqueness to visible light provide
the explanation. I suggest that generally in science cause and
effect involve more than sequential occurrence; they involve relationship in a theory. That is, they are not simply matters of fact
(observation); instead, cause and effect are highly theory dependent.
It is only in the context of a theory that we can ascribe the relation in a manner such that physical law or process requires the substantial, implicative succession of effect from cause. And, in
spite of Russell's claim to the contrary, I believe that much of
theoretical work in physics has the goal of showing how a theory
does provide cause-effect explanation. Do we not see this in the
search for understanding of superconductivity, of nuclear fission,
or of periodic variation in luminosity of stars? And, we reject a
correlation between sun spots and economic cycles as probably spuri-
ous, because we know of no relating process, but accept a correlation between sun spots and terrestrial magnetic storms because there
is a plausible physical relationship.

   It must still be granted that one can find much in contemporary
physics that supports Russell's thesis of an equivalence of causality
to functional dependence. The Lorentz transformations, for example,
have now been used in physics for some 70 years, and yet physicists
show little disposition to ask   why   relative motion gives rise
to a mass-energy increase, or to a time-rate change in physical processes. Perhaps, they do not because there is no ready prospect of
an answer; it has been remarked that scientists (unlike philosophers)
have a pragmatic judgement about working only on questions which are
amenable to solution. If no cause-effect explanation is to be
reached, one simply accepts, as we do in any event for so much of
both descriptive statement and axiomatic principle in science.
Nonetheless, science does progress by extending its domain of understanding. In doing this it gives insight causally into what is intrinsically happening, in the terms of a theoretical description
which elucidates with its own concepts and premises.   It seems,
therefore, that Russell's functional dependence principle must be
modified and supplemented by the assertion that causality remains
a root concern of natural science.

   It is interesting to note that Russell's collaborator in the
writing of Principia Mathematica, Alfred North Whitehead, developed
a comprehensive "philosophy of organism" in which there is a pro-
nounced relation-in-being corresponding to cause-effect. He, however, has not had an influence, I judge, that is at all comparable
to that of Russell.

V. CAUSALITY AND QUANTUM THEORY

A. Already in Hume we see the beginning of an empirical tradition in which there is no firm ground for necessity of causation.
But in physics proper a parallel tradition established itself, as
indicated in the quotation from Laplace, to the effect that a strict
cuasal determinism obtains everywhere in nature. The great change
that has come in physicists' outlook with respect to causality is
that forced by the quantum theory; and, because physics is the fundamental natural science, the change has deeply influenced ideas a-
bout nature among scientists generally, and in humanistic circles too.
The most apparent way that quantum mechanics entails indeterminism is in the inexactitude of position and momentum coordinates,
q_i and p_i. If future states of a system are to be predicted, presumably we must know the q_i's and p_i's for the present state of the
system. But we cannot know those simultaneously at any given point;
hence, the future of the system is unpredictable.
We state the quantum-theory indeterminism in a more perceptive
way, however, if we refer to the "projection postulate"; or to the
"non-rational" transition from a superposition of states, for a system,
to the system eigenstate which is observed. Nothing in quantum
theory tells us what particular eigenstate a system will enter, except that a probability measure may be calculated for a given state.
Hence, individual events, on the quantum level, are not within a
pattern of mathematical description or prediction, and there is an
inherent element of chance in nature. Or, we can say that there is
an abundance of events which are not in an invariant cause-effect
pattern.
In Russell's terms, the individual events of quantum physics
cannot even be fitted into relationships of functional dependence
(again, except statistical). Hume, I judge, would have been de-
lighted to have this vindication of his assertion that there is no
necessary relation in nature between cause and effect; his statement
that what seems such a connection is in fact only in the mind appears
now as a sharp and profound insight, and a score for the merit of
philosophical analysis. Aristotle, who was never faced with a
triumphant science based on a doctrine of universal and necessary
efficient causation, could probably have readily assimilated quantum indeterminism into his four causes. As stated, however, his
doctrine seems to assume that some cause is always acting, even
when luck or chance (both of which he discusses) seems to be present.
B. Most physicists now accept a lack of strict causal relationship between events in the natural world. There are also, however,
active groups both in philosophy and in physics who maintain, against
the Hume-Russell tradition, that there is an intrinsic determination
of every event, at every level of nature. For those who press this
principle, the eigenstate of an observation in quantum physics is
determined by prior states, even though quantum theory does not
today tell us what are the variables through which this determination is achieved.

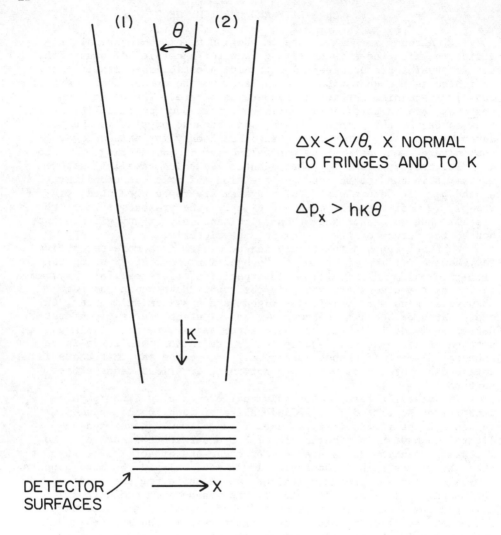

$\Delta x < \lambda / \theta$, X NORMAL TO FRINGES AND TO K

$\Delta p_x > hк\theta$

DETECTOR SURFACES $\longrightarrow$ X

Figure 1. Dual Source Intraphotonic Interference.
A simplified diagram of the Mandel-Pfleegor experiment: lasers (1) and (2) together send only one
photon into the apparatus at a time. An observed
difference in photon count on alternate detector
surfaces gives the evidence for interference.

It is not my intent to enter into a discussion of deterministic theories in current quantum-theory literature. I do, however, want to discuss one experiment which has not, I think, been publicized as much as it deserves to be, and which does, I think, give remarkably strong evidence against an underlying, deterministic agent that could prescribe each individual event on the quantum level. This is the demonstration, by Pfleegor and Mandel,[16] of interference effects by single photons from two different sources.

The experimenters used two independently operated laser beams, inclined to each other at a small angle θ and with each beam so attenuated that generally there would only be one photon at a time, from either beam, in the apparatus. Yet, they found that the momentary interference characteristic of incoherent sources did occur in the region of their detector surfaces. But if only one of the lasers operated, there was no interference at all.

Mandel and Pfleegor point out that their results can be understood in the following way. The fringes are in planes normal to the propagation vector $\underline{k}$ of the beams, $|\underline{k}| = 1/\lambda$, with an orientation such that photons must be localized within a distance $\Delta x < \lambda/\theta$ if fringe maximum is to be distinguished from minimum (Fig. 1). For, recalling that θ is the angle between wave fronts from the two lasers, we see that we must have $\Delta x \cdot \theta < \lambda$. But also, by the Uncertainty Principle, $\Delta p_x \geq h/\Delta x$. Hence, $\Delta p_x > h\theta/\lambda = h\theta k$. With a momentum uncertainty $\Delta p_x > h\theta k$, the photon may come from anywhere within an angle θ. So, for an observer at the detector surface there can be no stipulation (if we are to have interference) of which laser was the source of the photon, and it must be regarded as being in a superposition of two states, one for emission from each laser. (From a wave point of view, it is of course just such a superposition of wave states that leads to the interference.)

If in nature it actually were determinate as to which laser source emits the photon, we could not have the intra-photonic interference that is observed. The inability to identify the source, required by the Heisenberg Principle, therefore reflects a genuine uncertainty in nature. Any circumvention of the uncertainty, as by a hidden variable theory, would be counter to the Mandel-Pfleegor results. Hence, we can draw the conclusion that quantum physics supports an elemental indeterminism, on the level of the small particles of nature.

C. Quantum theory has also brought with it a possibility for an even more radical change in our conception of causality. Hume (and Kant) saw causal necessity as being a contribution of the mind, but the events (objects) of the natural world as being independent of the mind. But in quantum theory we have a subjectivism in that, a) a person can know an actual event, an eigenstate, only through an observation-interaction with it; b) the manifestation of the observed event (state) is to some degree controlled by the observer in his disposition of macroscopic objects (e.g., laboratory apparatus), there previously being, generally, a superposition of many states for the observed entity. This element of dependence of physical state on the observer can hardly be questioned, and constitutes a genuine subjectivist turn in natural philosophy; it is, however, a limited subjectivism, bearing only on the microscopic level of nature.

One interpretation of quantum theory, essentially proposed by J. von Neumann, and in recent years notably amplified by E.P. Wigner, prescribes that there be no eigenstate unless there is an interaction-observation with a conscious being. Otherwise, we must regard all states, whether for microscopic or macroscopic objects, as being distributed superpositions, governed rigorously by the quantum-mechanical laws of state functions. Since indeterminism enters with the passage to an eigenstate, and on this interpretation the locus of individual events of nature is in the consciousness of man (and other living beings?), consciousness then has an immediate relationship to relaxation of rigorous determinism. An even more general implication is that we lose independence from awareness for physical objects, on all levels of magnitude. Wigner suggests, therefore, that it may no longer be useful to maintain the hypothesis of an impersonal, universal reality.[17]

The Wigner approach is appealing in its opening of an intrinsic role for consciousness in physical processes; one can hope that progress in relating sentience to natural phenomena generally might result. On the other hand, the extension of the "Superposition × observation → eigenstate" principle to all objects can hardly be accepted uncritically; for, we have not found on the macroscopic level those superposition and interference phenomena which required a modified epistemology for microscopic physics. I shall not pursue the question further, except to comment that quantum theory apparently gives a possibility for yet another dramatic shift in thinking about causality.

VI. CONCLUSION

Our historical survey has brought up four general questions about causality:
1. Determinism - Indeterminism.
2. Role of causation; what is involved beyond invariant sequence of functional dependence? Is there elucidation in terms of physical process, so that causation is the essential element in scientific understanding?
3. Necessity of temporal succession, before-after, in association with cause-effect?
4. Degree of subjectivism in causality; how much is in nature and how much do we as knowers or scientists introduce?

The answers to all of these questions are strongly dependent, I suggest, on the physical theory that is accepted and established. Causality is not an a priori principle, or set of principles, but is rather a general characteristic of scientific knowledge. As such, it partly sets the form of science; but also, the history of causality amply demonstrates that the content of science responds back as a determinant of causality principles. The interplay continues today and very likely will in the future, as long as science is vigorously extending itself.

REFERENCES

1. My discussion is based on sec. 3 of Book B as given in Aristotle's Physics, trans. by H.G. Apostle, Indiana University Press, Bloomington, 1969.
2. ibid., Sec. 3, lines 18-24.
3. H. Butterfield, The Origins of Modern Science, Macmillan, New York, 1951, p. 5.
4. H. Butterfield, ibid., p. 9 et seq.
5. P.S. Laplace, A Philosophical Essay on Probabilities, trans. by F.W. Truscott and F.L. Emory, Dover Publications, New York, 1951, p. 4.
6. B.K. Bhattacharya, Causality in Science and Philosophy, Sanskrit Pustak Bhandar, Calcutta, 1969, p. 65.
7. The discussion of Hume will be based on his An Enquiry Concerning Human Understanding and selections from his A Treatise of Human Nature, Open Court Pub. Co., La Salle, Illinois, 1946.
8. ibid., Treatise of Human Nature, Book I, sec. XIV.
9. The presentation of Russell's philosophy of causality will be based primarily on his essay, "On the Notion of Cause", in Mysticism and Logic, W.N. Norton, New York, 1929. Although the essay was first published in 1912-1913 (Proc. Aristotelian Society, XIII, 1) Russell endorses it in the Preface to this volume as still representing a point of view which he regards as correct.
10. ibid., p. 180.
11. ibid., pp. 192-194.
12. ibid., pp. 195-196.
13. J.A. Wheeler and R.P. Feynman, Rev. Mod. Physics, 21, 425 (1949).
14. Mary B. Hesse has given a detailed discussion and criticism in her Forces and Fields, Nelson, Edinburgh, 1961, pp. 279-289.
15. R. Schlegel, Time and the Physical World, Dover, New York, 1968, pp. 17-20, 62-67.
16. R.L. Pfleegor and L. Mandel, Phys. Rev., 159, 1084 (1967); Physics Letters, 24A, 766 (1967). A discussion on a less technical level is given by R. Schlegel, Synthese, 21, 65 (1970).
17. Symmetries and Reflections: Scientific Essays of Eugene P. Wigner, Indiana University Press, Bloomington and London, 1967, p. 194.

DISCUSSION

Editor's Note: The following was edited by both the editor and the speaker from tapes of the question and answer period which followed the talk.

SAPERSTEIN: The impact that Newton and Laplace make is a differential formulation of nature. The functional relations are differential equations. Later on you get the integral formulation of nature, Hamilton or you go back even before Hamilton. Do you have any idea as to which caused which, did the philosophers cause the physicists to look at the integral approach or was this done independently of reading Hume, for example?

A. As far as I know I would guess that it was done independently of the philosophers. I think of something like Hamilton's principle; I think that that came through physics. Of course you can go way back: there is a Zeno, not of Zeno's paradox but another Zeno who gave us specular reflection on the basis, I believe, of some kind of a Fermat principle argument so these have come all along; but I would say this--I think this point's worth making--that in general, there is an equivalence always between the differential formulation and the integral formulation. Say you use Fermat's principle to derive Snell's law. You can also equivalently get it from differential equations.

SAPERSTEIN: The point is that it seems that the differential point of view and the success of it under Newton had an enormous impact on the philosophers and just about everybody at the time. One wonders whether the equivalent integral approaches damped down that impact, had their own impact.

A. I would think that certainly, yes, the differential approach as it were was the dominant one, carrying people along, I would certainly think so.

FADNER: While I'm perfectly willing to accept as a mode of operation the assertion that the first point you make about determinism and indeterminism is in favor of indeterminism, I really can't go along with the idea that it has been shown very well that this is true. For example, your case with the two laser beams; I don't think that at all shows that nature doesn't know where the photon came from. I think that I could make a very good argument quite the contrary that nature knows perfectly well where it came from. That not only does it know where it came from but it also knows that there is a second laser in which you are building up metastable states that are preparing, so to speak, to give off a photon of a certain frequency. And all we have to assert is that photons are mighty big or have some trailing or preceding part of the wave packet and I just can't go along with your conclusion that that's what nature knows. It seems to me what you can conclude is that the way we have set up this experiment and possibly the only way we can set up experiments is so that we don't know where that photon came from.

A. I don't think that I have any comment to make on what you say. I thank you.

NORDTVEDT: Would you like to comment on the Everett-Wheeler attempt to bring quantum mechanics back into more of an objective spirit, to downgrade the subjective element?

A. Of course this is the most exotic of all the interpretations
that we have. This in a word is, you know, the interpretation in
which we assume that we never do go from the superposition to an
individual eigenstate. But rather, say I could see a photon above
this line or below this line, these are the two concerned substates.
What happens is that I divide so that there are two me's. One of
them will see it up here and the other will see it down here. In
other words, the universe is continually subdividing. And in this
way, one maintains a complete, formal, rational description of
nature. I think there are all kinds of arguments against it. I
mean in terms of the probability of the existence of the many real
worlds, the unlikelihood that we have these divisions. I think
it's an amusing intellectual exercise.

MEYER: Might biology not correct Aristotle in that the mother as
well as the father is found to be the efficient cause of the baby.
But if time can be eliminated in fact as well as in our minds, in
the sense that Russell wants to eliminate it so that the future
is the cause of past as well as the past being the cause of the
present, then there should be instances where a son such as myself,
could be the efficient cause of my mother or my father.

A. Well, I think that if you accept that causality is not tempor-
ally directed, you'll have to say that's true. That it is all one
complete pattern. And it is just as reasonable I suppose to say
that you're the cause of your parents, as they are of you. I don't
like this, but I think there is something else involved here
though without going to that extreme. The final cause of Aristotle
has always been taken to involve a certain teleology that in some
sense does seem to involve that in some way what the future state
is, determines what is at present. And I think biologists accept
that in some way this is true. Somehow, not in any, let's say,
'metaphysical' way, but the very fact that the nature of the future
organism somehow determines the growth processes of the child.

DRESDEN: I have a comment and a question. My comment, and I will
speak as a physicist and perhaps because of it or perhaps inspite
of it I never understand philosophy very clearly, but it seems to
me that the notion of causality, independent of its philosophical
underpinning, is to be judged on its utility in terms of predicting
what is going to happen. That is my comment. Now my question is
really the following: When one looks at the phenomena then there
is always a certain number of events of a single part. You can get
certain sequences. But causality has something to do with the or-
ganization of these type of events. However, it seems to me that
in this there are two things which one has to look at. One is the
scale on which these events are being considered. And the scale is
in a sense a macroscopic scale or a microscopic scale which leaves
the question of determinism or indeterminism time reversibility or
non-time reversibility these are crucially scale dependent, so that
is point one then. Point two is that not only are they scale dependent but also in some sense as you have said they become theory
dependent. But it is really this notion of theory which has a

18

great deal to do with the fact of the predictability. The reason,
for example, why I prefer your explanation of the two laser experiments, is because if you had told me the set-up and if I had enough
energy and were smart enough, I could have computed how it would
come out. The alternate proposal that this is not so, which is of
course, logically always feasible, does not have that same predictive power. So I will now be able to comment first of all on the
fact that there is no strong theory dependence, but also a very
strong scale dependence and therefore I'm unsure and always I
feel most uncomfortable when somehow a clean sweep is made about
the whole set of principles, without identifying the scale.

A. Well, I certainly agree about the theory dependence and I think
if I understand you, that I agree on the scale dependence because I
don't believe that quantum theory for example, impugnes the work of
celestial mechanists in calculating when the next eclipse is going
to be. We, in fact, know of course by the correspondence principle
that when we have ensembles of many systems, quantum theory goes
over to a large scale approximate determinism. I think this point
is also very much relevent to things like the Wigner proposal. Do
we, when we no longer have interference effects, still have to
think that we have superpositions and the choice has to be made to
find the eigenstate.

DRESDEN: Also it has a great deal to do with the biological situation because again there are perfectly causal, well I don't want
to use that word, well-defined stages, well-defined developmental
patterns but again to look at those and to identify them as repeatable, for that one again has to look at the proper scale.

SAPERSTEIN: I wonder if perhaps another way to look at it going
back to Aristotle, we know that in the Greek view a non-manipulative view of the universe--as one   way of thinking, you can
give up causality and say 'well differential or integral are equivalent, I don't care which.' Hume can get away with that. But
the fact remains that a major part of Western thinking is a manipulative approach to the universe. And that manipulative approach
requires some time ordering and 'if I do this now, then that will
happen then'. It may be that the integral and differential equations are analytically equivalent but psychologically there is an
enormous difference, and I think the scale, the rapid philosophical
history you have given, I think has ignored this manipulative approach which I think is characteristic certainly of the physicists.

A. I think the very fact that we have had the limitations to efficient cause in much of modern physics is an indication of that,
because the efficient cause is the one that immediately gives rise
say to the effect.

SUDARSHAN: I would like to make two comments--one a general comment and one a more specific one. The first one is that in supplementing your remarks for the history of the thing, in the Indian tradition there are three main schools which may be distinguished. (I hope there aren't any experts on India Philosophy here.) The first one that with this variation is perhaps the most common, the most akin, with the modern physical point of view in which the observation is that the seeker after truth observes the universe and the universe includes himself in all the activities and in this one it is a passive non-manipulative observation and the Nirvana, or release, or the attainment of knowledge is in seeing the lawfulness of the universe, the apparent coming together of things. There is a succession of events of configurations and there are no things which go from one place to another place just as you mentioned about Hume, but there are only tendencies, Sanskara, inertia which goes along from one point.... The second tradition is that of the Jaines which breaks up in a certain sense the material universe and the spiritual universe in which you gain attainment of wisdom in the seeing of what is not yourself, in seeing that there are a whole lot of things. The individual entity has very little to do with all that is happening and when this attachment falls off then essentially the universe continues to be causal but you are no longer involved in its goings on. And thirdly, the Hindu tradition is probably the most difficult to summarize because different people have different views on it. But the non-humanistic or the monistic tradition of Shankar and his predecessors talk about the fact that in principle all things start and end in the person who knows, in the observer. That in fact, even when we talk about the knowledge of an object, the object of which you think you have the knowledge, itself rises and falls in the observer, and that all laws start and end with the observer. The only purpose of lawfulness is so that your concepts about the things can be ordered. (I could go on to quote Sanskrit to impress you but that probably would not quite do the job.) So that in a certain sense the views of Wigner for example are probably closest to the Hindu view of existence and information that in fact there are no objects and no matter and no experience without a theory, and no theory without the observations, and that these two cannot be disentangled and in fact must be seen in the perspective of the person who is knowing. You may for convenience, detach part of your experiences and theory and then call this material "universe", but in fact the question is really one of perspective and the change in the perspective is therefore the change in the observer's point of view. The more specific statement I wanted to make consisted in the fact that in a certain sense in talking about Aristotle and physics, much of Aristotle's physics was wrong but on the other hand much of Aristotle's views were right as you yourself pointed out. There are certain views of Aristotle which one sort of makes fun of all the time for example Aristotle said that planets move in circles because they are the perfect figures. But we find no hesitation in talking about various particles in high energy physics and then say we must have a particle at

1870 MeV, because of the fact that if you don't have this one then this particular Regge trajectory would not go through this place, or that we put these things together because these form a representation of the group SU3, or we must have two polarization states. So that the whole general notion of group theory or algebraic schemes of conservation laws of the general geometric approach to things seems to be in a sense a reflection of the injection of the global point of view of Aristotle which has not been sufficiently emphasized. It is a very different statement to make that the movement is according to the inverse square law than to say that you have a circular figure and that it is a nice thing to have a circular figure. Most people seem to think that these value judgements are very bad, except when they make them.

DRESDEN: Just when somebody else makes them.

SUDARSHAN: Right!

DENMAN: Speaking about Russell's association of causality with functional dependence, since most functions are readily invertible, that would force him into saying, for example, that night causes day as well as day causes night.

A. That's right, yes.

DENMAN: If that's all you associate with causality is a functional role, then when you invert the function, you are inverting the causal relationship.

SCHLEGEL: So, he had to go along with night causes day.

FADNER: That's true, its a one to one mapping.

NEWTON: I just wanted to make a remark about George Sudarshan's last statement. I very much agree with you. I think that that's perfectly true. We are arguing in physics nowadays also very often from beauty, mathematical simplicity and things which are, essentially esthetic criteria. But I think the examples that you give, point to me to a certain deficiency in the state of physics right now because I think that at least many physicists would feel that that is a temporary state of affairs that we would not explain things any more deeply than that. In other words, that we will not stay on the level of group theory forever, one would hope. The level of group theory is extremely beautiful and abstract and colorful and yet does not, I think, satisfy one's craving for a mechanism or something. That kind of thing is missing, I think, in physics at this stage.

DUFF: It seems to me that the Wigner interpretation then gives us the idea that quantum interference experiments are detectors of intellect. Should it not then be possible to devise experiments to examine what intellect is, to define it's quality? Is that a possibility?

A. Well, I don't really know what you mean by saying quantum interference experiments are detectors of intellect. Certainly the
situation is required that we have interference if we are going to
have consciousness having the role of giving us the eigenstate. I
would just make this general remark. I think Wigner's ideas are
extremely important and I think that in them we may be seeing some
beginning of how we can understand consciousness, according to the
quantum theory, because it is the first time in physical science as
far as I know that we have an explicit role for consciousness.

SUDARSHAN: I must point out, that there is a very small, very
simple, very elegant book by Schroedinger called "My View of the
World", much of Wigner's views appear for obvious reasons (there
is a good treatment), in which Schroedinger goes on to point out
that   it is not only true of quantum mechanics but in fact
the notion of consciousness and the perceiving and the coming in
appears also in classical physics and in practically every branch.
So, I strongly recommend that this enters the subject stream. It
is a very simple book; I even have it with me.

A. Yes, I know that book, it's very good. It seems to me Wigner
has added a different sort of thing, but it certainly is a very
good book.

CAUSALITY AND RELATIVISTIC DYNAMICS

Peter Havas

Department of Physics, Temple University, Philadelphia, Pa.   19122

ABSTRACT

It is pointed out that one must distinguish between causality as a general property of some physical theories expressing that the time development of <u>closed</u> systems is determined if the initial state is suitably specified, and specific cause-effect relations arising from the action of an outside agent on an <u>open</u> system. The connection between the space-time structure of the special theory of relativity and its causality requirements is discussed briefly, and some common misunderstandings are clarified. The consequences of these requirements for classical relativistic theories of systems of interacting mass points are discussed; it is argued that they do not impose any restrictions on the description of closed systems, but only of open ones, since it is only the latter for which the concept of causal anomaly is physically meaningful. Thus no restrictions are imposed on the possible existence of particles moving faster than light (tachyons) within closed systems, but the possibility of purposeful production of and experimentation with tachyons must be excluded. Similarly, for closed systems, particles are not restricted to interactions describable by fields; indeed, at least to order c^{-2}, any field-theoretical interaction acting over intervals in or on the future light cone of the source (commonly mislabeled "causal") is equivalent to an interaction (similarly mislabeled "noncausal") acting over a space-like interval.

INTRODUCTION

Arguments based on very general considerations, such as "symmetry" or "causality", can be very powerful tools in science. If used correctly, they can provide a shortcut to important conclusions, and avoid fruitless experimental or theoretical searches; if used carelessly, they can stop exploration of areas which might have yielded valuable new insights or discoveries. Einstein's theory of relativity brought about important changes in the outlook of physics on causality. Although their implications, in broad outline, were understood early, some of these implications were soon forgotten, and others misunderstood, with particularly unfortunate consequences for the development of relativistic dynamics.[1] In the following, I shall mainly discuss two cases involving the causality requirements of the special theory of relativity, first the consequences of these requirements for the problem of particles moving faster than light ("tachyons"), and second some recent developments in the relativistic dynamics of interacting particles, developments which were long delayed by a misunderstanding of these requirements.

Before we can elaborate on the causality requirements of the theory of relativity, we must briefly consider the prerelativistic situation and the usual connotation of "causality" in physics.[2] We are all familiar with the everyday usage of the words "cause" and "effect"; it frequently implies the interference by an outside agent (whether human or not), the "cause", with a system, which then experiences the "effect" of this interference. When we talk of the principle of causality in physics, however, we usually do not think of specific cause-effect relations or of deliberate intervention in a system, but in terms of theories which allow (at least in principle) the calculation of the future state of the system under consideration from data specified at a time t_o. No specific reference to "cause" or "effects" is needed, customary, or useful, but it is understood that all the phenomena (or variables) which can influence the system have been taken into account in the initial specification, i.e., that the system is closed. Conversely, if the system is open, i.e. if interference by an outside agent is allowed, no prediction of the future state of the system from its present one is possible. If the interference is arbitrary, no scientific statement at all can be made; if it is specified as a definite function of time instead, the state of the system at times $t > t_o$ may still be calculated, but it is not a function of the state at t_o alone.

A clear distinction between open and closed systems is essential if confusion is to be avoided in discussing the problem of causality in physics, within a prerelativistic as well as in a relativistic framework, and this distinction will be fundamental in the considerarations presented here.

CAUSALITY IN NEWTONIAN MECHANICS

The prototype of a successful causal theory is Newton's mechanics of mass points. In this theory the acceleration of each particle at a given instant times its mass is equated to the total force acting on this particle; the force between any two particles is a function of the positions (and possibly the velocities) of the particles at the instant considered. If the force laws are known and the "initial data" are specified, the positions of the particles are determined as functions of t for all $t > t_o$ or at least as long as the system can be considered as closed. The forces of Newtonian point mechanics are instantaneous action-at-a-distance forces; no reference to any medium between the particles is made in the theory, nor to any mechanism of transmission, any gradual spreading of "effects" due to the particles. For such a system the number of initial data required to specify the future development is finite; for n mass points we need 6n numbers (the 3n components of position and the 3n components of the velocities at $t = t_o$).

The future development of the system is determined from a knowledge of the force laws and the initial data even if the system was not closed at $t < t_o$: for the behavior of a Newtonian system it is irrelevant how it reached the state of t_o, whether by natural development (i.e. having been closed at all $t < t_o$), or by the influence of phenomena no longer active at t_o. This allows experimental verification of the predictions of Newtonian mechanics not simply by observation of an initial state as found in nature and later comparison of the results of calculation and observation, but by deliberate setting up to a particular initial state.

But if the system was indeed closed at all $t < t_o$, its past behavior can also be calculated for all earlier times; thus, if we chose to call our initial configuration the "cause" of all the other configurations determined by it mathematically, we would have "effects" at earlier as well as at later times than the "cause", which is not a particularly fortunate choice of language. In the theoretical description of a closed system, everything is interconnected; it is purely arbitrary to single out one aspect of the system as "the cause" of another.

On the other hand, Newtonian mechanics can also deal with open systems, i.e. systems in which there are "external" forces not described by any laws, such as a force exerted by a man on a body. Here the body alone is "the system". The man is not part of it; only the force exerted by him enters the equations describing the system. This force is arbitrary, and could thus be called "the cause" of the acceleration of the body. While it must be specified as a function of time to allow calculation of the motion of the body, there is nothing in the physics of the situation requiring the force to remain the function originally specified; it could at any time be changed to a different function, a "cause" which would have the "effect" of a different motion.

THE THEORY OF RELATIVITY

In Newtonian physics the existence of an absolute time and of an absolute meaning of simultaneity had been taken for granted. As Einstein realize, this implied the existence of signals of infinite velocity allowing the instantaneous, unambiguous synchronization of distant clocks, whereas he, guided by experiment, took as a basic postulate of the theory of relativity the existence of a maximum signal velocity in nature. He then showed that the existence of this limiting signal velocity implied that the concept of simultaneity of distant events was not absolute, but involved an element of definition; the most convenient way of establishing what is meant by "the same time" at widely separated points is by means of the fastest signal available, namely light in empty space. The same definition must be adopted for all inertial systems to allow the development of the theory on the basis of Einstein's two postulates: I. When properly formulated, the law of physics are of the same form in all inertial systems; II. In all such systems, the velocity of light in empty space has the same value c.

The existence of a maximum signal velocity implies a very un-expected result, which is of fundamental importance for the problem of causality. In Newtonian physics an experimenter using suitable equipment could in principle communicate with (send a signal to) any point in space, no matter how distant, in an arbitrarily short (even zero) time interval, i.e. operating at time t_0, he could in principle influence any event anywhere at all times $t \geqslant t_0$ and simi-larly all events occuring anywhere at all times $t \leqslant t_0$ could influ-ence him. According to the theory of relativity, however, he could not send any signal faster than c and thus in a time $t - t_0$ he could only reach points at a distance $\leqslant c(t - t_0)$; conversely, he could only be reached by those signals emitted at an earlier time t' which originated at distances $\leqslant c(t_0 - t')$. This is represented graphically in Fig. 1 (where the third spatial dimension is omit-ted). Only the region within or on the forward part of the cone shown there (the "future light cone") can be reached by signals from (causally connected with) its apex (x_0, y_0, t_0) and only signals emitted from within or on the backward part of the cone (the "past light cone") can reach the apex. The entire region outside the cone

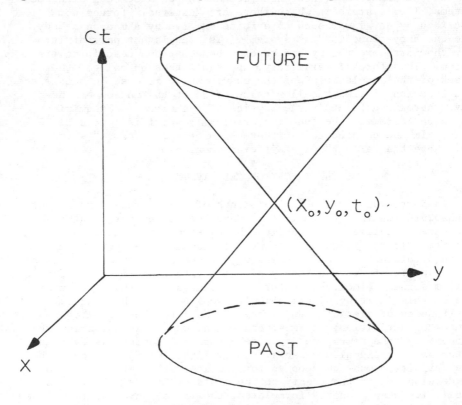

Fig. 1. The light cone.

can not be causally connected with the apex; by a suitable coordinate transformation any point within this region can be made simultaneous with it, and thus in a sense the entire region outside the light cone constitutes the "present" for the observer at the apex. In Newtonian physics this region collapses into a plane $t = t_o$ separating the past and the future.

Points outside the light cone are said to have "space-like" separation from the apex, those inside the cone have "time-like" separation, and those on the cone have "null" or "light-like" separation. Thus it is the region with space-like separation which can not be causally connected with a point.

CAUSALITY, CAUSAL CONNECTION, CAUSAL ANOMALIES

The appearance of a region which can not be causally connected with a point, i.e. which can not be influenced by anything happening at that point and can not influence it, is a fundamental new feature of the theory of relativity.[3] But what is its significance for our discussion of causality?

We have just used the words "causally connected" and "influence" rather loosely. But we <u>had</u> to use them loosely, in the naive, anthropomorphic sense of "cause" and "effect", in the sense in which the "effects" of a "cause" can be actually traced, because it is only in this sense that signals can be recognized, and it is the necessity of the use of signals to establish the meaning of simultaneity of distant events which is at the very basis of the theory of relativity. We were not talking about causality in the sense of predictability in a closed system because, as we discussed earlier, in such a system we can <u>not</u> identify one part of a phenomenon as the cause of another, trace the spreading of effects, or recognize signals.

The very concept of signals is based on a macroscopic interference with a system, as it is this interference which allows recognition of physical phenomena as signals, i.e. transmission of information. A signal must therefore involve an element of irreversibility, as has been stressed particularly by Mehlberg[4] and more recently by Terletskii[5]. Only for such phenomena does the theory of relativity restrict the speed of transmission to be less than c. No limitation is imposed on speeds not suitable for signal transmission; examples are the phase velocity of an electromagnetic wave in a wave guide or in a region of anomalous dispersion, or the relative velocity of two bodies moving in opposite directions with speeds close to c relative to an inertial system[3].

The reason that--once the theory of relativity is accepted--one must exclude the possibility of signals traveling faster than c is that otherwise it would be possible to construct a causal anomaly, as first recognized by Einstein[6]. This causal anomaly is usually simply stated as "an effect preceding a cause", but to avoid a purely semantic argument this must be spelled out more clearly. The existence of a causal anomaly will be taken as equivalent to a logical contradiction of the following type: An experimenter turns on the light at t_o; this light activates a mechanism which interposes an obstacle between the experimenter and the switch <u>before</u> t_o, so that he can not turn on the light at t_o. Or, more drastically--the light activates a mechanism which kills the experimenter's grandfather a century before t_o, so that there is no experimenter to turn on the light!

The requirements for the (supposed) existence of such a causal anomaly are twofold: First, it must be possible to produce a "cause" at a point A in space at time t_o which has an "effect" at the <u>same</u> point A at an earlier time; second, this effect must be of sufficient strength to interfere with the original cause, and re-peatable--otherwise the occurrence could be dismissed as a mass hallucination.

TACHYONS AND CAUSAL ANOMALIES

To establish the possibility of an effect preceding a cause by means of superluminal signals, it is simplest to consider a "Minkowski diagram" with a single spatial dimension (Fig. 2). We consider two intertial systems S (with coordinates x and ct) and S' (with coordinates x' and ct') moving with a relative speed $v < c$.

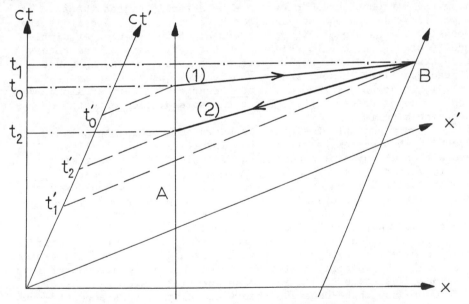

Fig. 2. Minkowski diagram for two successive signals leading to a causal anomaly.

Observer A, at rest in S, produces a signal at t_o traveling with speed $V > c$ to the right (labeled 1), which arrives at the point

B, at rest in S', at time $t_1 > t_o$ (as measured in S). There it
instantaneously activates a signal emitted at t_1' traveling to the
left with speed $V' > c$ (labeled 2), arriving at A at time $t_2' > t_1'$
(as measured in S'). However, this time t_2, as measured in S, is
earlier than t_o. Clearly $t_o - t_2$ can be made as large as desired
by increasing the distance between A and B. Furthermore, no
observer is needed at B; the reception of the first signal and
emission of the second could be accomplished by a mechanism. Thus
to avoid a causal anomaly, we must disallow the possibility of a
macroscopic signal by the experimenter at A.

In less elaborate form, a similar argument was used by Einstein
to exclude the possibility of a "method of transmission where the
effect obtained would precede the cause (possibly accompanied by
an act of will)".[6] More specifically, it was used by Laue[7] to
disprove the possibility of the existence of a rigid body. With
specific reference to signals, it was pointed out by Terletskii[5]
that it was not the existence of particles moving with velocities
greater than c which was excluded by the theory of relativity, but
that "forbidden only is the process in which the emission of such
particles is systematically repeated and associated with an increase
in entropy of the radiator", a necessary condition for the transmission of information.

In spite of this, in recent years it has been suggested by a
number of authors that "tachyons" might exist and might be produced
experimentally.[8,9] Most of their effort has gone into a reinter-
pretation of the world lines of tachyons 1 and 2 of Fig. 2. While
world line 1 proceeds into the future as observed from S, and 2 into
the future as observed from S', they each proceed into the past
as seen from S' and S, respectively (and also have negative energy,
a point which can not be discussed here). It has been shown[8,9]
that by means of a "switching principle" it is possible to describe
these world lines as having positive energy and proceeding into
the future. In short, what has been convincingly demonstrated is
that if we only observe world lines of tachyons, it is possible to
describe these world lines in a consistent manner not violating
logic or "causality". But what has been achieved is simply a consistent description of tachyons forming part of a closed system; the
problems of causality violation can only arise in open systems, however, and these are not problems of description, but of self-con-
tradictory phenomena. While there have been a number of papers
objecting to various aspects of the arguments advanced by the
tachyon advocates[9,10], this distinction has perhaps not been sufficiently emphasized. Within closed systems the existence of tachyons
is indeed fully compatible with the causality requirements of the
theory of relativity. However, purposeful production is not, at
least not to the extent that it is repeatable and may lead to macroscopic effects, which would make it suitable for signaling.[11] If it
is not repeatable, i.e. the production is completely random, it is
independent of the intent of the experimenter and thus unsuitable
for signaling. To avoid macroscopic effects, it is not enough to
exclude the possibility of production of a macroscopic beam of tach-

yons; one must also exclude the possibility of triggering of other
macroscopic phenomena by tachyons (except again in a completely random fashion). These limitations rule out the possibility of design-
ing any experiments involving tachyons, and only allow chance observation.

Thus the theory of relativity imposes serious restrictions on
the processes allowed in a mechanics of open systems. We now turn
to the question of the mechanics of closed systems.

THE PROBLEM OF RELATIVISTIC POINT MECHANICS

We consider the motion of n mass points. In Newtonian mechanics, the motion is specified by giving the positions and velocities
at time t_0, since we are dealing with instantaneous action-at-dis-
tance forces. However, the system of Newton's equations of motion
and force laws does not conform to the relativity postulates, and
thus such a description is unacceptable. Nevertheless, we can ask
ourselves whether it is possible to find a new set of equations of
motion and of force laws, which does conform to the relativity
postulates, and which still allows us to predict the motion of the
particles from a set of data (not necessarily the same as in
Newtonian mechanics) given at $t = t_0$ in any inertial system.

This problem is the natural generalization of the Newtonian
problem of n bodies. Nevertheless, the question was apparently
not posed in this form when the theory of relativity was created,
and not investigated until 1949, due to a lack of distinction between
open and closed systems, and a consequent erroneous belief that the
causality requirements of the theory of relativity imposed the same
restrictions on closed systems as on open ones. This led to the
belief that a force law which depends on the simultaneous positions
of two particles corresponds to an instantaneous spreading of
"effects"; by this same misunderstanding it was also frequently
assumed that the force exerted by one particle on another could only
manifest itself as a retarded action, i.e. only within its future
light cone. However, forces as such have no direct physical significance; they are only auxiliary quantities entering the laws of
motion, the physically observable quantities being at best the positions of the particles as functions of time. Thus the problem we
are concerned with is that of particles left to move freely under
their mutual actions, and the possibility of specifying this motion
by a set of data at time t_0. This is not the same problem as starting the motion at t_0 with the specified initial data, which is
indeed incompatible with the requirements of relativity; the determination of the world lines by arbitrary starting data given at
spatially separated, but simultaneous, points would indeed imply that
the arbitrarily set data at the position of one of the particles
influence the motion of the other particles at the same instant, and
thus correspond to transmission of information with unlimited speed.
But if the motion has been proceeding freely at earlier times, the
setting of "arbitrary" initial data at t_0 is only a mathematical
arbitrariness; we do not imply that we physically set these data at

t_o, but only ask the question how the motion would proceed in a case in which these data are observed at t_o.

Such a distinction between the case of free motion of a closed system with a set of data existing at t_o, and a motion started at t_o with the same initial data (which requires that the system was open up to that time) does not have to be made in Newtonian mechanics. It is essential for relativity, however.

An alternative to the necessity of introducing this distinction would be to require that the effects of the particular values of the initial data should make themselves felt only after times excluding signals exceeding c. The customary procedure is to specify data on the full surface $t = t_o$, rather than only at the points of intersection of the world lines, thus introducing infinitely many additional ("field") degrees of freedom, and to require that the laws determining the time development of the variables (which now are functions given at all points of space-time rather than on isolated world lines) should only involve near action, i.e. action between neighboring points, rather than action at a distance (in analogy to Newtonian continuum mechanics rather than point mechanics). Maxwell's electrodynamics was a ready-made example of such a theory, and thus it came to be almost universally considered as the model of a relativistic theory.

However, for closed systems there is no need to pattern a theory after this model, and there is no logical reason preventing the development of a relativistic point mechanics with Newtonian causality, i.e. a theory in which the state of the particles can be specified by 6n data.

There are three general approaches to a relativistic dynamics of mass points patterned after the Newtonian model of forces acting at a distance.[12] One is to replace the Newtonian Galilei-invariant force laws by (explicitly or implicitly) Lorentz-invariant ones, an approach taken by Poincaré[13] even before the special theory of relativity was created, but then neglected for half a century. This approach has been developed recently by a number of authors[14]; unfortunately, while it has been shown that force laws exist for which Newtonian initial data are sufficient to determine the motion, the proof either is valid for two particles only, or the force laws so far discovered are extremely special and of little physical interest.

The second approach is to replace the Galilei-invariant canonical formalism by a Lorentz-invariant one. This approach was initiated by Dirac[15]; one of the formalisms he suggested has the advantage of dealing with variables at a single time, as in Newtonian dynamics. Dirac himself was unable to demonstrate that this formalism is not empty. Bakamjian and Thomas[16] were able to show that it allows a description of interacting particles provided that one renounces the concept of Lorentz-invariant world lines for these particles; on the other hand, it has since been shown through various "no-interaction" theorems (first developed by Currie[17]) that maintaining this concept within the canonical formalism excludes the possibility of interaction[18].

The third approach is to base the equations of motion on Lorentz-invariant variational principles. Such variational principles depending on the four-dimensional separation of the particles and on the product of their four-velocities were first introduced for electrodynamics by Fokker[19], and developed by various authors[20]. Principles including a general dependence on other relativistic two-body invariants depending on the separation and velocities of the particles were introduced and studied recently[21-23]; the resulting equations of motion form a subclass of the equations considered earlier by Havas and Plebanski[14,12].

In the following we shall be concerned with particle systems described by such Lorentz-invariant variational principles. For a particular form of interaction, such as Fokker's[19], it is possible to introduce "adjunct fields",[24] and thus it is possible to some extent to compare the properties of interactions of particles describable only by action-at-a-distance theories and of other interactions which may also be describable by field theories with supposedly different "causal" properties.

VARIATIONAL PRINCIPLES AND TYPES OF INTERACTIONS

We shall be concerned with a classical dynamical system consisting of interacting point particles. The notation used to describe this system will be identical to that used in Ref. 22. As long as we are dealing with a fully relativistic system it is convenient to describe its motion in a four-space with coordinates $x^\mu (\mu = 0,1,2,3)$, where x^0 is the time coordinate. Repetition of a Greek index implies summation over this range. The metric of this space is

$$\eta_{\mu\nu} = 0 \quad \text{if} \quad \mu \neq \nu,$$
$$\eta_{00} = 1, \quad \eta_{11} = \eta_{22} = \eta_{33} = -\frac{1}{c^2}. \tag{1}$$

The world line of the j-th particle, with coordinates z_j^μ, will be parametrized by the proper time τ_j, where

$$d\tau_j = \left(\eta_{\mu\nu} \, dz_j^\mu \, dz_j^\nu \right)^{\frac{1}{2}}. \tag{2}$$

We define a four-velocity $v_j^\mu(\tau_j)$ and a four-acceleration $a_j^\mu(\tau_j)$ by

$$v_j^\mu(\tau_j) \equiv \frac{dz_j^\mu}{d\tau_j}, \qquad a_j^\mu(\tau_j) \equiv \frac{d^2 z_j^\mu}{d\tau_j^2}, \tag{3}$$

so that

$$v_j = \left(v_j^\mu \, v_{j\mu} \right)^{\frac{1}{2}} = 1, \qquad v_j^\mu \, a_{j\mu} = 0. \tag{4}$$

On the other hand the space of Newtonian mechanics is the usual three-dimensional space with Euclidian geometry. In this space the path of the j-th particle $\vec{r}_j(t)$ is parametrized by the absolute time t; the three-velocity $\vec{v}_j(t)$ and three-acceleration $\vec{a}_j(t)$ are defined as

$$\vec{v}_j(t) \equiv \frac{d\vec{r}_j}{dt} \quad , \qquad \vec{a}_j(t) \equiv \frac{d^2\vec{r}_j}{dt^2} \quad . \tag{5}$$

This space will be used also to describe a system whose equations of motion are only approximately Lorentz-invariant.

We now postulate a Lorentz-invariant variational principle

$$\delta I = 0 \quad , \qquad I = I_1 + I_2 \quad , \tag{6a}$$

where I_1 is defined by

$$I_1 \equiv -\sum_{i<j}\sum g_i g_j \iint_{-\infty}^{\infty} d\tau_i \, d\tau_j \, \Lambda_{ij}(s_{ij}{}^\Lambda, v_i{}^\Lambda, v_j{}^\Lambda) \, , \tag{6b}$$

$$s_{ij}{}^\Lambda \equiv z_i{}^\Lambda(\tau_i) - z_j{}^\Lambda(\tau_j) \, ,$$

and I_2 will be considered below. The interaction is characterized by n "coupling constants" (introduced in analogy to the constants used in electro- and mesodynamics) and by $\tfrac{1}{2}n(n-1)$ possibly distinct functions Λ_{ij} (not necessarily symmetric in i and j). A further generalization can be introduced by considering different types of interactions between a given pair of particles, each similarly characterized; this will not be done explicitly at this stage, but it should be understood that the final result could be generalized to any number of types of interactions as noted. Each Λ_{ij} is assumed to be invariant under the infinitesimal transformations of the proper orthochronous Lorentz group and to depend only on the positions and velocities of the particles. Thus it can be a function only of the two-body invariants of the group, which will be given later.

The variations of the coordinates of each world line $z_j{}^\mu$ are not independent, since Eq. (4) implies

$$v_j{}^\Lambda \, \delta v_{j\mu} = 0. \tag{7}$$

Thus the term δI_2 is needed to maintain these conditions. It is defined by

$$\delta I_2 \equiv -c^2 \sum_i \int_{-\infty}^{\infty} d\tau_i \, M_i(\tau_i) \, v_i{}^\Lambda \delta v_{i\mu} \, , \tag{6c}$$

where the M_i's are Lagrangian multipliers, and a factor of c^2 has been introduced so that they have dimensions of mass. It is not necessary to define I_2 explicitly in this formulation. The minus signs in (6b,c) are chosen so that (6) can have the usual Newtonian limit. The M_i's can be determined from the Euler-Lagrange equations, as shown in Ref. 22, but will not be needed here.

The "usual Newtonian limit" means a variational principle

$$\delta I = 0, \quad I = \int_{-\infty}^{\infty} dt \; L[\vec{r_i}(t), \vec{v_i}(t)], \quad i = 1 \dots n, \quad (8a)$$

with

$$L = T - V, \qquad T = \frac{1}{2} \sum_i m_i \vec{v_i}^2,$$
$$V = \sum_{i<j} \sum g_i g_j V_{ij}(r_{ij}), \qquad r_{ij} \equiv |\vec{r_i}(t) - \vec{r_j}(t)|. \quad (8b)$$

The g_i's are constants characteristic of the interaction and are included for convenience in comparing the relativistic case to (8); the m_i's are the masses of the particles. The potential energy $g_i g_j V_{ij}(r_{ij})$ depends only on the instantaneous interparticle separation. Forces derivable from such potential energies are static and central. The variational principle (8) is invariant under the infinitesimal transformations of the Galilei group (up to a divergence, i.e. a total time derivative). However, it by no means represents the most general Galilei-invariant variational principle[21]. Furthermore, it does not even yield the most general Newtonian point mechanics, nor is it the only possible limit of Lorentz-invariant variational principles[22]. However, here we restrict ourselves only to such interactions Λ_{ij} which lead to (8).

There are four independent two-body invariants which can be formed from the four-coordinates and four-velocities of two particles. If a Newtonian limit (8) of the relativistic variational principle (6) is desired, an appropriate set of invariants to be used as arguments of Λ_{ij} in Eq. (6b) is

$$\sigma_{ij} = c^2 \eta_{\mu\nu} s_{ij}^{\;\mu} s_{ij}^{\;\nu} = c^2 (t_i - t_j)^2 - |\vec{r_i}(t_i) - \vec{r_j}(t_j)|^2$$
$$\equiv c^2 t_{ij}^2 - \vec{r_{ij}}^2(t_i, t_j),$$
$$\omega_{ij} = v_i^{\;\mu} v_{j\mu} = \gamma_i \gamma_j \left[1 - \frac{\vec{v_i}(t_i) \cdot \vec{v_j}(t_j)}{c^2} \right],$$
$$\gamma_i \equiv \left[1 - \frac{\vec{v_i}^2(t_i)}{c^2} \right]^{-\frac{1}{2}}, \qquad (9)$$
$$\chi_{ij} = c\, v_{i\mu} s_{ij}^{\;\mu} = -\gamma_i \left[c\, t_{ij} - \frac{1}{2} \vec{v_i}(t_i) \cdot \vec{r_{ij}}(t_i, t_j) \right],$$
$$\xi_{ij} = c\, v_{j\mu} s_{ij}^{\;\mu} = +\gamma_j \left[c\, t_{ij} - \frac{1}{2} \vec{v_j}(t_j) \cdot \vec{r_{ij}}(t_i, t_j) \right],$$

which has a static Newtonian limit, as shown in Ref. 22. Then, as also shown there, we can reduce our Lorentz-invariant variational principle to a variational principle of the form (8a), but Lorentz-invariant (up to a divergence) to order v^2/c^2, with

$$L \equiv L_2 - V + I_{PN}, \qquad (10a)$$

where

$$L_2 = -c^2 \sum_i m_i \left(1 - \frac{1}{2} \frac{\vec{v}_i^{\,2}}{c^2} - \frac{1}{8} \frac{\vec{v}_i^{\,4}}{c^4}\right), \quad \text{(10b)}$$

V is the Newtonian potential energy as given in (8b), and the "post-Newtonian" interaction I_{PN} is given by

$$I_{PN} \equiv \frac{1}{2c^2} \sum_{i<j} \sum g_i g_j \left\{ \vec{v}_i \cdot \vec{v}_j \; V_{ij}(r_{ij}) \right.$$
$$- \vec{v}_i \cdot \vec{r}_{ij} \; \vec{v}_j \cdot \vec{r}_{ij} \; \frac{1}{r_{ij}} \frac{dV_{ij}}{dr_{ij}} + (\vec{v}_i - \vec{v}_j)^2 [V_{ij}(r_{ij}) + X_{ij}(r_{ij})]$$
$$+ [(\vec{v}_i - \vec{v}_j)\cdot \vec{r}_{ij}]^2 \; Y_{ij}(r_{ij}) + (\vec{v}_i^{\,2} - \vec{v}_i \cdot \vec{v}_j) \; W_{ij}(r_{ij})$$
$$\left. - [(\vec{v}_i \cdot \vec{r}_{ij})^2 - \vec{v}_i \cdot \vec{r}_{ij} \; \vec{v}_j \cdot \vec{r}_{ij}] \frac{1}{r_{ij}} \frac{dW_{ij}}{dr_{ij}} \right\}. \quad \text{(10c)}$$

Here V_{ij} is the Newtonian potential, which can be defined from the relativistic function Λ_{ij} by a procedure described in Ref. 22. Similarly, X_{ij}, Y_{ij}, and W_{ij} are functions which can be defined from Λ_{ij} and in general are independent of V_{ij}.

The theory described by Eqs. (10a)-(10c) can be put into canonical form. Indeed, it has been shown[25] to be the most general theory, to order c^{-2}, which can be constructed starting from a Newtonian canonical theory. It thus represents the most general dynamics which could have been obtained (to order c^{-2}) by means of the second approach discussed earlier. In addition, it satisfies the world line condition to order c^{-2}, and the difficulties discussed earlier do not arise to this order.

For reasons to be discussed below, we shall be concerned in the following only with Λ_{ij}'s symmetric in i and j. Then W_{ij} vanishes and I_{PN} reduces to

$$I_{PN} = \frac{1}{2c^2} \sum_{i<j} \sum g_i g_j \left\{ \vec{v}_i \cdot \vec{v}_j \; V_{ij}(r_{ij}) \right.$$
$$- \vec{v}_i \cdot \vec{r}_{ij} \; \vec{v}_j \cdot \vec{r}_{ij} \; \frac{1}{r_{ij}} \frac{dV_{ij}}{dr_{ij}} \quad \text{(10d)}$$
$$\left. + (\vec{v}_i - \vec{v}_j)^2 [V_{ij}(r_{ij}) + X_{ij}(r_{ij})] + [(\vec{v}_i - \vec{v}_j)\cdot \vec{r}_{ij}]^2 \; Y_{ij}(r_{ij}) \right\}.$$

As noted above, there may be more than one type of interaction in the interaction term I_1 of (6) or (10). This would lead to appropriate summations over the various types of interaction within V as well as I_{PN} in Eqs. (10), a possibility which will be exploited in the following.

The variational principles (6) or (10) are formulated in terms of direct particle interactions, without any reference to possible underlying field theories. Only those interactions for which "adjunct fields" can be defined in terms of the particle variables could possibly arise from an underlying field theory, as discussed in Ref. 22, where it was shown that only interaction terms I_1 of the form

$$I_1^{(\ell m)} = - \sum_i \sum_{i<j} g_i g_j c \iint_{-\infty}^{\infty} d\tau_i \, d\tau_j \; v_i^{1-\ell-m} \, v_j^{1-\ell-m}$$

$$\times \; \omega_{ij}^{\ell} \, \chi_{ij}^{m} \, f_{ij}^{m} \, \phi_{ij}^{(\ell m)}(\sigma_{ij}) \tag{11}$$

(and their linear combinations) allow the definition of fields adjunct to each particle. Here ℓ and m are non-negative integers, and the $\phi_{ij}^{(\ell m)}$ are arbitrary functions of their arguments. The approximately relativistic interaction following from (10) is

$$I^{(\ell m)} = \frac{1}{2c^2} \sum_i \sum_{i<j} g_i g_j \left\{ \left[(1-\ell-m)(\vec{v}_i - \vec{v}_j)^2 \right. \right.$$

$$\left. + \; \vec{v}_i \cdot \vec{v}_j \right] V_{ij}^{(\ell m)}(r_{ij})$$

$$- \left[\vec{v}_i \cdot \vec{r}_{ij} \; \vec{v}_j \cdot \vec{r}_{ij} - \frac{m(1-m)}{2m-1} \left((\vec{v}_i - \vec{v}_j) \cdot \vec{r}_{ij} \right)^2 \right] \frac{1}{r_{ij}} \frac{dV_{ij}^{(\ell m)}}{dr_{ij}} \right\} , \tag{12a}$$

where

$$V_{ij}^{(\ell m)}(r_{ij}) = (-1)^m \int_{-\infty}^{\infty} d\zeta \; \zeta^{2m} \, \phi_{ij}^{(\ell m)}(\zeta^2 - r_{ij}^2). \tag{12b}$$

Clearly both the exact interaction (11) and the approximate one (12) are symmetric in the particle variables. Thus no combination of interactions of the type (12) can match the general direct particle interaction (10c). It is for this reason that we restrict our attention immediately to the symmetric interaction (10d).

Only interactions of the form (11) allow the definition of adjunct fields. Conversely, for all known special relativistic linear field theories for which the equations of motion have been discussed in the literature, the time-symmetric equations of motion of simple point particles have been shown to be derivable from a variational principle with interactions of the form (11), with $m = 0$ and $\phi_{ij}^{(\ell m)}$ equal to the time-symmetric Green functions of the field equations. Interactions of the form (11) with $m > 0$ were shown in Ref. 23 to arise similarly for appropriate linear field theories, with field equations which are partial differential equations of order $4m + 2$ (which of course may be a consequence of lower order equations). However, it has not been shown that this is true for all conceivable linear field theories. This problem will not be treated here, and thus, in the following, the term "field-theory related interactions" will be used in the sense of "particle interactions allowing the definition of an adjunct field theory" rather than in the probably equivalent sense of "particle interactions following from a linear field theory by use of time-symmetric Green functions."

One of the most important characteristics of direct particle interactions of the type (6b) is that interactions may exist between pairs of particles with space-like separation. Although there exist linear field theories with Green functions depending on space-like separations, most interactions considered in the usual field theories derive from Green functions depending on time-like or null separations. Therefore, we shall in the following pay special attention to interactions with time-like or null separations, i.e. to relativistic interactions of the form (11) with $\sigma_{ij} \geqslant 0$, and the corresponding approximately relativistic interactions (12a,b) with

$$\xi^2 - \vec{r}_{ij}^{\,2} \geqslant 0. \tag{12c}$$

Therefore the term "field-theory related interaction" will in the following imply the condition (12c) as well.

EQUIVALENCE OF INTERACTIONS OF DIFFERENT TYPES

Obviously, in general the Lorentz-invariant direct particle interaction (6b) or its linear combinations, even if it is symmetric in the particle variables, can not be mathematically equivalent to the field-theory related interaction (11) or its linear combinations. However, this is not true for the corresponding (symmetric) approximately relativistic interactions (10d) and (12); rather, it has been established[23] that any field-theory related interaction of class C^∞ of the form (12), which approximate a Lorentz-invariant field-theory related interaction of the form (11) with time-like or null particle separations to order c^{-2}, can be compounded in an infinity of ways from interactions of the form (10d) arising from a Lorentz-invariant non-field-theory related interaction of the form (6b), but which depends on the invariants (9) such that the particle separations are space-like.

What, if any, are the implications of this result for the question of causality? If the concepts of causality as applicable here are properly understood, there are none: we are dealing with a closed system, and causality only means that proper choice of initial data (to order c^{-2}, just as in the Newtonian limit, simply the positions and momenta) determines the future (and past) motion uniquely. However, if, as is far too common, one believes that interactions "propagate" and thus must act over time-like or null intervals only, the result is instructive because it shows that all such interactions can, to order c^{-2}, be equivalently described by interactions acting over space-like intervals; furthermore to order c^{-2}, all time-symmetric interactions over time-like or null intervals are equivalent to retarded interactions, i.e. interactions acting over intervals in or on the future light cone of the source only, and thus it is also these interactions commonly mislabeled "causal" which are equivalent to the interactions similarly mislabeled "noncausal"[26].

THE INITIAL VALUE PROBLEM

Little is known about the more fundamental question of the proper choice of the initial data for the exact (explicitly or implicitly) Lorentz-invariant equations of motion describing an n-particle system. All possibilities are open. Newtonian specification is implicit in the second approach discussed earlier, based on a canonical formalism, if one requires a single-time formalism (which leads to the difficulties discussed before). Without this restriction one obtains a 6n-parameter family of solutions; this, however, is not necessarily equivalent to Newtonian specifications. The possibility of Newtonian specification in the first approach was discussed earlier; for general Lorentz-invariant equations of motion the multiplicity of solutions is not known. For such equations following from the variational principles considered in the third approach, this multiplicity is known only in two cases. One is the strictly one-dimensional case [with the Green function of the one-dimensional wave equation inserted in (6)] for two particles considered by Staruszkiewicz[27]. There exists a four-parameter family of solutions; however, these are non-analytic, and for certain initial positions and velocities there exists an infinity of solutions which can not be reduced by further specification of accelerations or their derivatives (which indeed must all vanish if any solution is to exist). The other case is Fokker's two-particle system with unsymmetrical interactions[28], one charge acting with retarded field, the other with advanced field; this yields a 12-parameter family of solutions. For Fokker's time-symmetric interaction, it has recently been established (first through perturbation methods[29], then through consideration of exact solutions[30]) that for two charges there may exist an infinity of solutions for given initial positions and velocities. However, no general theorem are known even for this case, much less for the case of general interactions.

CONCLUDING REMARKS

In our brief survey of various aspects of causality in relativity we had to make a fundamental distinction between open and closed systems. Causality in the sense in which it is incorporated in Newtonian point mechanics can also be achieved in relativistic point mechanics, but while in Newtonian mechanics this required only that the system be closed from a time t_0 for which its state is specified up to a time t_1 until which prediction is desired, the postulates of relativity require that such a system be closed for all time. This is a serious restriction, but it is not a unique one; Einstein's theory of general relativity, which is a field rather than a particle theory, must impose a similar requirement[1]. However, other aspects of both these theories pose difficulties the consequences of which have not been fully explored. These diffi-

culties are connected with the physical requirements any theory has
to satisfy, in particular that any verification of the predictions
of a theory is possible only if experimentation is permitted, i.e.
if the system is not rigorously closed at all times.

One possibility of escaping from this difficulty is to apply the
theory to macroscopic systems for which observation does not
significantly disturb the system (just as in the analogous case of
Newtonian celestial mechanics). Another might be to renounce
detailed verification[1]. Such a verification is impossible in
practice anyhow even in Newtonian mechanics if we are dealing
with a microscopic theory. It might be that we shall not be able
to use the full richness of results apparently contained in the
theory, but only some macroscopic averages (as in statistical
mechanics) or some statements of asymptotic behavior (as is sometimes assumed in elementary particle physics).

The preceding exposition has been entirely restricted to classical theory. The conclusions arrived at in our discussion of
tachyons are not modified if quantum effects are taken into account.
The relativistic dynamics of mass points clearly can be applied to
microscopic phenomena only within the framework of a quantum theory.
Such a theory has been developed by Bakamjian and Thomas[16] within
the framework of a canonical formalism[31].

Much work remains to be done both in the classical and the quantum domain. The development of physics has suffered from a one-
sided approach, which was based less on considerations of experimental findings than on a misunderstanding of the implications of
the postulates of relativity. As experiments reveal more and more
about the behavior of particles at ultrarelativistic energies,
without any corresponding advances in understanding the phenomena,
it is imparative that theoretical research should explore the full
richness of formalisms which can be fitted into the framework provided by these postulates.

FOOTNOTES

1. Various aspects of the relation of relativity and causality
   are discussed by P. Havas in Proceedings of the 1964 International Congress of Logic, Methodology and Philosophy of
   Science (North-Holland Publ. Co., Amsterdam, 1965), p. 347,
   and in Synthese 18, 75(1968).
2. For an elementary discussion of causality from the point of
   view of a physicist see V. F. Lenzen, Causality in Natural
   Science (C. C. Thomas Publisher, Springfield, 1954); for an
   extensive discussion of causality by physicists turned philosophers see P. Frank, Das Kausalgesetz und seine Grenzen
   (Springer-Verlag, Wien, 1932), and M. Bunge, Causality (Harvard
   University Press, Cambridge, 1959).
3. The space-time structure of the special theory of relativity
   sketched here is discussed in all textbooks of this theory;
   a particularly detailed account is given in H. Arzeliès,

La cinématique relativiste (Gauthier-Villars, Paris, 1955) [English translation: Relativistic Kinematics (Pergamon Press, New York, 1966)]. For a concise discussion of some of the implications of the theory of relativity for causality see H. Weyl, The Open World (Yale University Press, New Haven, 1937).

4. H. Mehlberg, Studia Philosophica 1, 119 (1935) and 2,1 (1937).

5. Ya. P. Terletskii, Dokl. Akad. Nauk SSSR 133, 329 (1960) [English translation: Soviet Physics-Doklady 5, 782 (1960)]; Paradoxes in the Theory of Relativity (Plenum Press, New York, (1968).

6. A. Einstein, Ann. Phys. 23, 371 (1907).

7. M. Laue, Phys. Zeits. 12, 85 (1911).

8. O. M. P. Bilaniuk, V. K. Deshpande, and E. C. G. Sudarshan, Am. J. Phys. 30, 718 (1962); G. Feinberg, Phys. Rev. 159, 1089 (1967).

9. For a general review see O.-M. Bilaniuk and E. C. G. Sudarshan, Phys. Today 22, No. 5, 43 (1969) and the subsequent discussion, Phys. Today 22, No. 12, 47 (1969).

10. E.g., R. G. Newton, Phys. Rev. 162, 1274 (1967) and Science 167, 1569 (1970); W. B. Rolnick, Phys. Rev. 183, 1105 (1969).

11. While the possibility of signaling by means of tachyons has been suggested by O.-M. Bilaniuk and E. C. G. Sudarshan, Nature 223, 385 (1969) and by E. C. G. Sudarshan in Symposia on Theoretical Physics and Mathematics, ed. by A. Ramakrishnan (Plenum Press, New York, 1970), Vol. 10, p. 129, it appears from their reply in the second paper of Ref. 9 that they no longer maintain this suggestion.

12. The development up to 1964 is discussed by P. Havas in Statistical Mechanics of Equilibrium and Non-equilibrium, ed. by J. Meixner (North-Holland Publ. Co., Amsterdam, 1965), p. 1.

13. H. Poincaré, Rend. Circ. Mat. Palermo 31, 1 (1906); similar attempts, mostly applying only to particles moving with constant velocity, were undertaken at the time by Sommerfeld and others, but were rapidly abandoned in favor of field-theoretical approaches.

14. P. Havas and J. Plebański, Bull. Am. Phys. Soc. 5, 433 (1961) (see also Ref. 12); H. Van Dam and E. P. Wigner, Phys. Rev. 138, B1576 (1965); D. G. Currie, Phys. Rev. 142, 817 (1965); R. N. Hill, J. Math. Phys. 8, 201 and 1756 (1967); L. Bel, Ann. Inst. H. Poincaré 12, 307 (1970) and 14, 189 (1971); Ph. Droz-Vincent, Phys. Scripta 2, 129 (1970).

15. P. A. M. Dirac, Rev. Mod. Phys. 21, 392 (1949). A review of this approach is given D. G. Currie and T. F. Jordan in Lectures in Theoretical Physics, ed. by W. E. Brittin and A. O. Barut (Gordon and Breach, New York, 1968), Vol. X-A, p. 91.

16. B. Bakamjian and L. H. Thomas, Phys. Rev. 92, 1300 (1953).

17. D. G. Currie, J. Math. Phys. 4, 1470 (1963); D. G. Currie, T. F. Jordan, and E. C. G. Sudarshan, Rev. Mod. Phys. 35, 350 (1963).

18. It has therefore been suggested that one should give up the identification of the canonical coordinates q_i with the physical

positions of the particles [E. H. Kerner, J. Math. Phys. 6, 1218 (1965); R. N. Hill, J. Math. Phys. 8, 1756 (1967); A. N. Beard and R. Fong, Phys. Rev. 182, 1397 (1967)]. However, then the principle of relativity becomes vacuous [A. Peres, Phys. Rev. Letters 27, 1666 (1971); 28; 392 (1971); this classical result also is implied by the quantum-mechanical study of R. Fong and J. Sucher, J. Math. Phys. 5, 456 (1964)]. Another modification of Currie's "world line condition" on the g_i was suggested recently by L. Bel, C. R. Acad. Sc. Paris 273, 405 (1971).

19. A. D. Fokker, Z. Physik 58, 386 (1929).
20. P. Havas, Phys. Rev. 87, 309 (1952); 91, 997 (1953); R. C. Majumdar, S. Gupta, and S. K. Trehan, Progr. Theor. Phys. (Kyoto) 12, 31 (1954); A. Katz, J. Math. Phys. 10, 1929 and 2215 (1969).
21. P. Havas, in Problems in the Foundations of Physics, ed. by M. Bunge (Springer, New York, 1971), p. 31; for a correction of some misprints see footnote 37 of Ref. 22.
22. H. W. Woodcock and P. Havas, Phys. Rev. D6, 3422 (1972).
23. F. Coester, P. Havas, and H. W. Woodcock (to be submitted to Phys. Rev. D shortly).
24. Introduced for Fokker's principle in electrodynamics by J. A. Wheeler and R. P. Feynman, Rev. Mod. Phys. 21, 425 (1949), and for the variational principles of mesodynamics by P. Havas Ref. 20.
25. J. Stachel and P. Havas (to be submitted to Phys. Rev. D shortly).
26. For an example of a quantum mechanical "noncausal" Hamiltonian equivalent to a "causal" one, yielding the same dispersion relation and scattering amplitude, see S. Gasiorowicz and M. A. Ruderman, Phys. Rev. 110, 261 (1958).
27. A. Staruszkiewicz, Am. J. Phys. 35, 437 (1967); see especially the Note added in Proof.
28. A. D. Fokker, Physica 9, 33 (1929).
29. C. M. Andersen and H. C. von Baeyer, Phys. Rev. D5, 802 (1972).
30. D.-C. Chern and P. Havas (to be submitted to Phys. Rev. D shortly).
31. The theory of "shadow states" presented by E. C. G. Sudarshan at this symposium also may be interpreted as a quantum theory of action at a distance.

DISCUSSION

Editor's Note: The following was edited by both the editor and the speaker from tapes of the question and answer period which followed the talk.

DRESDEN: I'd like to ask one question and maybe a second. You wrote down what I think was a very beautiful result connecting the results of the approximate relativistic particle interaction and a field theory. Now my first question is when you make these approximate calculations, then is it correct that in the results which you end up with, just in the part of the picture, these are no longer strictly Lorentz invariant, is that correct?

A. They are only Lorentz invariant up to $1/c^2$.

DRESDEN: No, but my question is that when you say you make the comparison with the field theory, do you make a simple, comparable approximation in the field theoretic calculations?

A. Sure. I take perfectly exact relativistic theories and compare them at the level of $1/c^2$.

DRESDEN: And the statement is: up to that level of comparison every field theoretic interaction can effectively be compounded from direct particle interactions.

A. Compounded, but not necessarily visa versa.

SAPERSTEIN: You make the distinction between open and closed systems and it seems to be a very convenient dichotomy. In one case you can talk about causality and in the other case you can't. Or as you say, it is not useful. But I wonder if it is a real distinction because whenever we talk about initial conditions, it seems to me at least implicit is the assumption that we can change the initial conditions. The initial conditions are ours to play around with just like your man pulling the box to decide not to pull for an hour, but to pull for 50 minutes. And he can do it, if you will, at his command. It seems to be the same thing, that that's what we assume about initial conditions. We can arbitrarily change the initial conditions. If that's the case, there really isn't any difference between those two, and I think this shows up in the fact that all your difficulties come in on how you specify initial conditions.

A. No, as I said at the very end, this is exactly the difference with Newtonian physics, that you are not free to play with the initial conditions. So, taking the theories at face value, all you can say is that if I happen to observe that such and such are the conditions at a certain time, I can go home and calculate the motion and then compare it with observation. But, I cannot go into the lab and set up these initial conditions and compare the motion with that calculation. It is not uniquely determined if I don't say that the system was closed earlier.

FADNER: In electrodynamics we have a situation that seems to me is somewhat similar to the tachyon situation in that we can by temporarily forgetting about the conservation of energy, receive a particle, say a photon or possibly a meson, before we send one out, and this doesn't bother us because we feel we can presumably upset conservation of energy for a short time. But when we

do this, we avoid having an infinite progression of such things by
truncating the number of Feynman diagrams that we are going to
allow ourselves to play with, and I think we convinced ourselves
that we know why we truncated it and that it is a reasonable
truncation. Now my question is--in the tachyon case, is it the
case that you have not come to any way to truncate the total number
of events in an analogous manner to what we do in electrodynamics?

A. Are you talking about two particles interacting or are you
talking just about the tachyons, as such?

FADNER: I was talking about the tachyon being a medium for the
interaction between two particles.

A. Well, there is no medium as such. That was the main point of
my discussion. I mean that in this picture there is no meaning in
saying that the interaction is transmitted, something is actually
happening in between. The only things which enter are quantities
which depend on the positions and velocities of the particles.
And everything else, while one can make pictures of it if one wants
to, I think this makes it more difficult. In this case I think
that the pictures confuse the issue, rather than clarifying it.

FADNER: You are separating space-like interactions from the concept of the tachyon, is that it?

A. Right.

STAPP: You said that you could carry these considerations over to
quantum theory, but it seems to me that in quantum theory you are
dealing basically with an open system rather than a closed system.
The very logic of quantum theory requires that the measurements
are treated in a certain way--are imposed as initial conditions and
the final measurements imposed from without. We translate these
boundary conditions into wavefunctions to calculate transition probabilities, so it seems that quantum theory is intrinsically an
open system, where you impose initial conditions and final conditions. Therefore it seems to me you are in trouble if you try
to use these tachyon or acausal type of interactions within the
framework of quantum theory.

A. First of all, I fully agree with you that the quantum theory in
principle is an open system and that's a big difference. I made
that statement in connection with the discussion of tachyons. What
I didn't say was that in quantum theory the big difference is that
you are not talking about individual events; if you are talking
about suitable averages, you can make all the statements I made
about signal production and not being able to set up macroscopic
situations in which you could influence the past, just like in
classical theory. An individual tachyon doesn't do any harm,
either in classical or in quantum theory. But you are absolutely
right as far as the second part is concerned, there is a question
of the interpretation, precisely.

SAPERSTEIN: Going back to this business with closed and openness, in classical physics when you said that in a closed system we just observe the situation right now and then you go back and predict. But isn't it true that not only must you observe the situation right now but the complete past of it? You must have the world line of whatever your are observing back to minus infinity before you are going to be able to predict the future?

A. Nobody knows. I mean it is ununderstood mathematics, in almost all cases. The question is if I give you the laws guiding this, and then ask what other things do I need to specify; is it positions and velocities, is it positions, velocities and accelerations, is it a finite stretch of the world line, infinite stretch of the world line, etc? The answer in general is not known. It is simply that the mathematics is not available.

SAPERSTEIN: You just pointed out that you could cut off Newtonian dynamics and you would get two different results, predictions, backward. But it would not be the same system, because somebody's turned it off. The same option presumably is available in the relativistic system.

A. No, what I said is that we are not allowed this cutting off because then you could use the way you set up the system to pro-pagate signals. For example, here (in Fig. 3) you have your ex-periment where you are going to shoot off particle one with a given velocity and then this will produce a certain effect, in-stantaneously, at the position P_2 of particle two; then a different initial velocity of particle one would produce a different motion of particle two, if indeed the initial conditions plus the laws would fully determine the motion, and thus you would have an in-stantaneous knowledge of what happened at position P_1. You have to exclude that possibility if you want a limit on signal velocities.

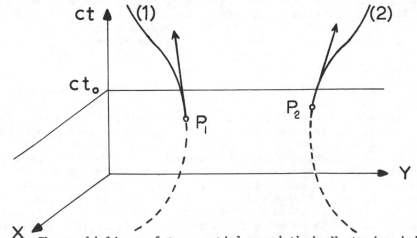

Fig. 3. The world lines of two particles and their Newtonian ini-tial data at time t_o.

GOBLE: Your example of a one-dimensional theory showed that for some theories it's not sufficient to specify conditions on a space-like surface. I wonder if that theory or other theories you know about rule out specifying conditions on the light cone, on any light cone?

A. No, for example, one example that I didn't mention for lack of time which is due to Fokker and has been reinvented several times within the last few years is that in electrodynamics, well in a sort of electrodynamics, he has two particles--one acting on two by means of the retarded field and two acting on one by means of an advanced field. Then what you have is precisely a one-to-one connection between points on the two world lines along their light cones. Therefore, specifying positions and velocities on the light cone is enough: that determines the motion.

NORDTVEDT: How do you do this for three or more particles?

A. That is a different question of course, because in more than one dimension particle two being on the light cone from one, and particle 3 on the light cone from two, doesn't put particle three on the light cone from one; I take it that this is what you meant. So that can't be done.

SCHLEGEL: In classical mechanics we help ourselves by reducing a two-body problem to a single body using a reduced mass, has this ploy been tried with the relativistic interaction?

A. Well, it cannot be done in general. That is a sad thing.

SCHLEGEL: What happens if you try to set up a reduced mass relativistically?

A. You just can't. You cannot do it if you are talking of Lorentz invariant interactions because then you cannot simply consider those two particles at the same time. The mathematics of going to the reduced mass just doesn't go through.

SUDARSHAN: I would like to make one remark. The remark is something which Buddha or Nagarjuna could have made, namely that if you know that everything is according to the law, then you cannot manipulate that news. If everything is purely observational and completely according to law you cannot set up initial conditions and you cannot observe, you cannot do anything. Therefore even in elementary classical mechanics, we have a very anomalous paradoxical situation of all physics. If you cannot measure, if you cannot set up experiments, there is no physics; but if there are no laws, then also there is no physics. If everything is completely according to law then you cannot set up an experiment because it's pre-ordained what you are going to do. This is not really only the relativistic mechanics or quantum mechanics or field theory or anything of this kind; we must have in a sense, the partially open manipulative universe. The notion of manipulative universe contains both closed and open parts. The question is related to this. When our friend Einstein and other people talk about signal velo-

46

cities did they define what was meant by signal or was this sort
of like is said about causality in elementary electrodynamics
books? Is it defined or undefined? Contained within it was there
the notion of cause and effect? Was the manner in which the
signal is being sent out relativistically invariant; is it taken
as an added assumption?

A. Well, it was. I don't remember the exact wording. But, it was
pointed out that this was an actual manipulative universe.

SUDARSHAN: The notion of what was the signal and what was being
sent, etc. This was required to be Lorentz invariant. That is,
one notion is of causality of some conditions at one time propagating onwards toward another set of conditions at another time. Another one is that this lawfullness, this obedience to a set of
laws, and therefore complete predictability, that it had to be
also arranged in such a fashion that what could be considered as a
manipulative signal, was to be interpreted as manipulative by
all observers.

A. Right, right. That was explicit. I fully agree, I said before
that one must allow some experimentation, and it seems to me that
the obvious example to my mind is the one I gave--a really macroscopic situation. One of the reasons I got interested in this was
precisely that I felt that there was something very strange that
you can talk about, say, celestial mechanics in the Newtonian sense
without any difficulty, and then if you want to go to special relativity, would you really have to throw everything out, and start
with an entirely different set of concepts, just to be able to describe the situation from the point of view of different inertial
frames. And it seems to me that there is no reason why one cannot
have a special relativistic macroscopic dynamics in exactly the
sense in which you take for granted in Newtonian physics that there
are situations in which the effects of the observational procedure
on the system to be observed are infinitesimal.

COOKE: Would you be willing to speculate that suppose all of the
particles were far apart in the past and they were all in straight
world lines, perhaps that would determine the future. Given some
kind of asymptotic straightness when they are all far apart.

A. You mean would that be enough to determine... I have come to
the conclusion that it is very dangerous to guess.

BELINFANTE: It seems to me natural that the initial condition in
a case like you just discussed would really be for all the space-
like surface to infinity.

A. Yes, but this is really a different problem. I did not discuss field theory. If you do field theory, of course, you also get
around this problem, but at the expense of having to specify infinitely many data. So the question here is always can you get
away with a finite specification as you can in Newtonian dynamics
of mass points? Of course, for continua, you also need an infinite
number of data in Newtonian physics, but for discrete particles

you don't and the question is--is there a possibility in special
relativity to do the same, and the answer is yes, in principle.

RELATIVITY AND THE ORDER OF CAUSE AND EFFECT IN TIME

Roger G. Newton
Indiana University, Bloomington, Indiana 47401

ABSTRACT

Signal velocities greater than that of light are well known to be ruled out by a combination of the theory of relativity and causality, because their existence would lead to the possibility of causal cycles. We exhibit the reason for this prohibition and analyze the relevant aspects of the notion of causality. We argue that the logical basis of the distinction between cause and effect is not their temporal order, but that it rests on the issue of control. We then examine the notion of control in this context.

It seems to me to be one of the great achievements of theoretical science to be able to conclude from statements of invariance, such as the relativity principle and the invariance of the speed of light, an absolute prohibition of the existence of any kind of signal that would propagate faster than light. The analysis of the exact basis of this conclusion should be able to cast much light on the way scientists think, or at least some of them. It should show how much of their thinking is purely deductive and how much is influenced by other considerations, or how many of their deductions are straight from explicitly stated premises, and how many are from tacit and frequently not very carefully examined presuppositions.

The three basic assumptions, the principle of relativity, the constancy of the speed of light, and the linearity of the transformation, as is well known, lead to the Lorentz transformation as the connection between the measurements of spatial coordinates and the time performed by two observers in uniform relative motion. From this transformation it follows that if it were possible to send a signal of any kind from one space-time point A to another, B, at a speed greater than that of light, that is, a superluminal signal, then there would always exist another possible observer in whose view the time of B is earlier than the time of A. Such an observer would therefore have to conclude either that the signal was sent from B to A, or that a signal was sent into the past. If it is possible to distinguish unambiguously between the sender and the receiver of a signal, which amounts to distinguishing between cause and effect, irrespective of which is the earlier, then the only conclusion he can draw is that a signal was sent into the past.

We may want to clarify what is meant by a signal. For our purposes it is quite sufficient to define it as anything that permits us to throw a switch at a space-time point B on command from a space-time point A. That, of course, raises the issue of control, about which I will say more later on.

This essentially very simple argument can be sharpened by using various moving observers, to form a causal cycle, as illustrated in

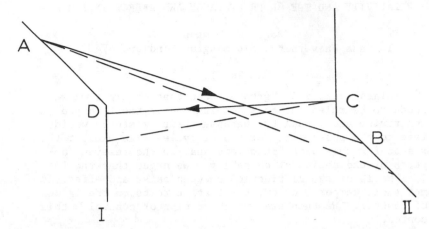

Fig. 1. A Causal Cycle. I and II are the world lines of two
          rockets. The dashed lines are lines of simultaneity as
          seen by both rockets. A superluminal signal connects
          A and B, and another, C and D. On both occasions ob-
          servers on both rockets agree who the sender and who
          the recipient is.

Figure 1. The heavy lines labeled I and II represent the world lines
of two rockets. Rocket I is first at rest and then moves at uniform
velocity, while rocket II first moves, at the same velocity as I,
and then comes to rest. (Alternately one may avoid all accelerations
and use four rockets. One then merely adds two signals as two
rockets pass one another closely.) The dashed lines are the lines
of simultaneity of the moving and the stationary rockets, respectively. Superluminal signals are sent from A to B and then from C
to D. It should be noted that in both instances the sender and
the receiver agree who is sending the signal and who is receiving
it; they are in each case at rest with respect to one another.

Thus, if superluminal signals exist (i.e., can be sent and detected)
then it is possible for an observer to send a signal into his own
past. He could then cause his own destruction and thus prevent the
signal from being sent in the first place. This impossible state
of affairs is often simply called a contradiction and as a result it
is said that the theory of relativity rules out the existence of
superluminal signals.

       There is, I think, more involved here than the theory of rela-
tivity. Let me make the point by a simple if somewhat weak analogy.
I want to send a given particle from A to B by giving it a specified
amount of kinetic energy. That may be possible. But if I am told
that the particle is to arrive at B less than 10 seconds after
leaving A, then it may be impossible. The impossibility arises from
the given boundary conditions. In the same way, the "contradictory"
causal cycle may be impossible to set up, not because superluminal
signals are impossible, but because of the stated boundary conditions

that involve the stipulated contradiction. That would mean that if I were to set up all the relevant apparatus for the "contradictory" cycle experiment and the time came for the final throwing of the switch, it simply wouldn't go!

The so-called contradiction has now been shifted to another plane. What bothers us now is not a logical contradiction, but a violation of experience or past experimental evidence. While there may be much everyday experience with switches or other gadgets not working a large part of the time, there is no experience at all with buttons not being pushable no matter how hard and often we replace them, and we would very much like to avoid giving such a blank excuse to the manufacturers of electronic gadgets.

The point I am trying to make is that in a scene that is tightly causally controlled both forward and backward, there is no difficulty with excluding the causal-cycle experiment if superluminal signals exist. The difficulty arises from our belief, that we <u>can push that button if we want to</u>. That is the essence of experimentation, and the great advantage physics has over astronomy or the social sciences. We can control and wiggle things, and see how other things wiggle in response. That's how we find out what causes what, and we are confident that this is a much more reliable way of studying cause-and-effect relations than mere passive observation of correlations.

What is the basic nature of the causal relation between events? Most of us no longer believe in efficient or necessary causes. I am a good Humean, and I say, it is simply <u>constant conjunction</u>. Except that we are now more sophisticated and we must weaken this to talk about <u>statistical correlations</u>.

Suppose, then, that we observe a correlation between a class of events A and a class of events B. How do we distinguish which is the cause and which is the effect? One possible answer to that question is to say that, <u>by definition</u>, the earlier of the two is the cause and the latter is the effect. In that case, if A causes B via a superluminal signal, another observer would have to conclude that B causes A, because the order in time of two points with a space-like separation is not invariant under Lorentz transformations. However, I will try to convince you that the definition of cause and effect is <u>not</u> based on their relative order in time.

Let me discuss a simple example, deliberately drawn so that we have no strong preconceptions preferring one interpretation over the other. (Or, perhaps more accurately, where we have equally strong preconceptions against both.)

Suppose that we performed an ESP experiment. In one room subject A throws a pair of dice at regular intervals, and in another room subject B writes down numbers between 2 and 12. Suppose now that after a long series of throws by A and writing by B we examine the records of both and we find a strong statistical correlation. If the experiment went on long enough and the correlation is strong enough then we would presumably have a <u>prima face</u> case for a causal relation between the two kinds of events. There arise then, however, two questions: 1) Is the statistical correlation really evidence

for a causal connection? Maybe both events are effects of a common
cause. 2) If there is a causal connection, which is the cause and
which, the effect? In other words, is it a case of clairvoyance,
or a case of psychokinesis?

I submit that there would be a simple way to answer both questions. We could have the dice throwing controlled by a more and
more remote device in whose randomness we have more and more faith.
If the correlation persisted we would have no choice but to conclude
that we have a case of clairvoyance. That conclusion would be independent of the time-order of the two events. We would have to
say that it was the throwing of the dice that caused subject B to
write down his numbers the way he did, even if we should also establish that he wrote them down <u>before</u> the dice were thrown!

It is the <u>control</u> over events that unambiguously distinguishes
a cause from the effect, not their order in time. There is, then,
nothing <u>logically</u> contradictory in the notion of an effect preceding its cause. What is violated by such a notion is not logic, but
experience or experimental evidence.

Suppose two classical particles move around under the influence
of a force between them. Does it make any sense to ask whether it
is A that causes B to move as it does, or whether it is B that
causes A to move the way it does? That, it would seem to me, is a
meaningless question. We may without fear of being wrong, assert
either. In a fully <u>time-reversal invariant situation</u>, whenever it
can be meaningfully asserted that A causes B, then it may be equally meaningfully asserted that B causes A. Only when the issue of
control enters, which involves a measure of non-invariance under
time reversal, can one meaningfully distinguish cause from effect.

Why is it that, Feynman-Wheeler electrodynamics notwithstand-
ing, in certain specific circumstances we may and do think of the
acceleration of a charge as <u>causing</u> the emission of radiation rather
than the other way around? We may do so because the definition of
radiation involves boundary conditions that introduce a failure of
time-reversal invariance into electrodynamics.

Let me go a little more deeply into the question of the relevance of time-reversal invariance. Let's first take a simple
classical situation where we can talk about constant conjunction.
If I assert that A causes B, I may be saying one of three things:
1) B happens if and only if A happens; 2) B happens whenever A
happens; that is, A is sufficient; 3) B happens only if A happens;
that is, A is necessary.

Suppose we have a situation in which A is a necessary and sufficient condition for B to occur. In that case I will find that if
I externally control A so that it happens at specific times t_1, t_2,
... (say, controlled by a random device), then B happens at specific
correlated times. But since A is necessary and sufficient for B
to occur, it will also be true that if I control B to occur at
specific times t_1, t_2, . . ., then A will happen at correlated
times. Hence A is the cause of B, <u>and</u> B is the cause of A. Therefore in the if-and-only-if situation there really is no way of distinguishing one cause-effect order from the other. After all, the
if-and-only-if relation is <u>symmetric</u>.

Let's look at the situation in which A is a <u>sufficient</u> condition for B to occur. Of course, we have to couple this with a
clause: "In specified circumstances." For example, I turn a switch
and the light goes on. If this happens once I may not conclude anything much. So I turn the switch many times, and every time the
light goes on. Then I conclude that my turning the switch <u>caused</u>
the light to go on. Of course this will be so only if the fuse is
not blown, and if the main switch is on, and if the bulb isn't burn-
ed out, etc. So these are the specified circumstances. But now
suppose suddenly the light goes on without my doing anything to
the switch. Would that invalidate my causal statement? No; I
would simply say that on that occasion something else caused the
light to go on. Maybe someone closed the circuit by another switch.
Therefore I conclude that if, under specified circumstances, A is a
sufficient condition for B, then A causes B.

On the other hand, take case 3): A is a necessary condition
for B. That is, B cannot happen without A. I control A and make
it happen many times. Sometimes B happens, sometimes not. But B
never happens without A. Then (in the context of constant conjunction) I would conclude that A is part of the "specified circumstances" and may be called a <u>contributing cause.</u> But I would not
call A the cause of B. On the other hand I would then call the
<u>non-occurrence</u> of A the cause of the <u>non-occurrence</u> of B. The
light went out because there was no electricity. This, of course,
confirms my previous analysis, because if A is necessary for B,
then non-A is sufficient for non-B.

So we conclude that a cause is, in specified circumstances, a
sufficient condition for its effect.

Now this simple picture becomes much more blurred when we have
to consider statistical correlations rather than constant conjunc-
tions. The distinction between the relevance of necessary and sufficient conditions is then no longer so clear. At the same time it
becomes, of course, more problematical to establish the causal relation. The social sciences and medicine are full of problems
connected with this.

Suppose one wants to establish a causal link between smoking
and lung cancer. What is found is a statistical correlation between the amount of smoking and the incidence of lung cancer. The
mor a person smokes, the more likely it is for him to get lung cancer. At the same time there are some people with lung cancer who
never smoked. One does not consider the existence of such cases as
invalidating the supposed causal relation. One says that in those
instances the cancer was caused by something else. Thus the cause
here too is a sufficient condition, but not necessary.

On the other hand, consider another medical problem. One
wants to establish if a specific micro-organism causes a particular
disease. In that case one will not only look for the bacillus in
people that have the disease, but one will observe if the person
comes down with the disease without the organism being found in his
body. In fact, if after we find the bacilli in someone's body, he
does not become ill, then one says the infection "didn't take,"
that is, the organism did not multiply for some reason; or the

patient was immune, etc. But if he gets sick and the bacillus is not found in his body, then one surely concludes that this germ does not cause this disease. So here we seem to have a case in which one takes the cause to be a <u>necessary</u> condition for the effect, but not sufficient.

I think this case can be reduced to our previous results by closer analysis. The argument that the organism did not "take" invokes a mechanism as a link between the original infection and the disease-causing action of the bacilli, which broke down. Hence, since one is not arguing that the original infection per se caused the disease, one should properly put the cause at the other end of the link, namely, the presence of the bacilli in large numbers. The other end of the chain is the disease. What is a disease? In one sense, it is a collection of symptoms; in another, it includes a "mechanism". Although one may well argue that in many cases the symptoms are so specific that a positive identification of the disease (including its "mechanism") is possible, I doubt that one can positively state that it is impossible for symptoms to simulate a disease, in the sense of including its mechanism. So we have severed both ends of the supposed necessary conditions: First to make the conditions both necessary and sufficient, by looking at the presence of large numbers of the bacilli in the body, rather than at the infection; then to make it unnecessary by allowing for the possibility that the symptoms are caused by something else. So we are back with a cause as a sufficient condition only.

In these cases we are dealing with observed statistical correlations, and we see that even then one can in principle distinguish between cause and effect, <u>if</u> the relation is present. There is, however, always the possibility that the observed correlation is not causal at all. In the case of smoking and lung cancer there is the possibility that both the urge to smoke and the cancer are caused by something else, that both are effects of a common cause. Similarly in the case of the disease and the bacilli. Possibly both are the effects of the presence of a chemical, or a "humor". Ultimately this question, which is present in any purely statistical correlation, can be resolved only in one of two ways: Either by the discovery of a mechanism, i.e., a detailed causal chain that leads from the alleged cause to the alleged effect; or by experimentation. The first way is really a reduction of the alleged causal relation to small causal steps for each of which the causal relation is known. The second way makes use of control over one end of the supposed causal connection, or its manipulation.

Let's go back now to that issue of <u>control</u>, and how it is exerted. The simplest thing is to invoke free will: I push that button when I want to! However, there being the possibility that free will is an illusion, this is not fool proof. I used random devices in the above arguments. One can use any other mechanism if only one can be confident that it is remote enough to be free of the causal chain under investigation. This is a matter of degree of course, and its judgement depends on theory. In our previous ESP experiment I could have the dice thrown or even <u>arranged</u> in

China and then the list of the results brought here and used a year later. I may feel confident that psychokinesis could not reach that far in space and time even if it did exist. But someone else might not be so confident. That is why a random device is useful. If it is well constructed, it will break to any desired degree, all causal chains to the past. It introduces an element of practical unpredictability even into a classical physical context. By means of a large number of instabilities it makes it for practical purposes impossible to continue a causal chain through it. If we let the times at which event A occurs be controlled by a well constructed roulette wheel, then even though we know that in principle the motion of the ball could be predicted from, and is causally controlled by, the initial conditions, we also know that for all practical purposes we can be confident that these times are not caused by something else. If the times of occurrence of B are correlated with those of A thus chosen, A must be the cause and not the effect.

The point of the introduction of a random device is always to break all causal chains as completely as possible. Of course, that cannot be done entirely, but one may do it to any desired degree of approximation. The judgment of that is based on previous experience and theory. There is no use pretending that it is absolute and independent of interpretations. That does not decrease its value or necessity.

Let us be more specific. There are two purposes of a random device such as a roulette wheel: The first is to make a definite decision, and the second, to break all causal chains. The second means that it must be essentially unpredictable, and the first, that it is essentially irreversible. I think it is useful to keep these two characteristics apart and to be clear about their origin.

The causal chain is broken in a random device by the use of instabilities. Small perturbations in the initial conditions will lead to large changes in the final state. A simple example would be to balance a hard sphere on a sharp edge. If I demand that it be set up in such a way that it must stay on the edge for a specified minimal length of time T then the larger T, the less predictable will be to which side the ball will fall when it finally does. That is so because the more closely the initial state approximates the exact unstable equilibrium state, the more sensitive the motion will be to small perturbations. By increasing T, I can make the outcome more and more unpredictable and hence I break the causal chain more and more completely. This is the simplest paradigm of a random device. It makes a definite yes-or-no decision. (I may want to put an upper limit on the time of balance. If it balances too long, I simply take it off.)

Now where does the irreversibility come in? Any decision among a finite number of choices can be reduced to a finite number of yes-no decisions. Hence we can always use a finite set of our ball-on-edge devices. But since the essential point is the instability with respect to small perturbations in the initial conditions, we must have a continuous variable there, or at last a variable that can take on a number of values that is very much larger than the final

two. As a result there has to be a <u>loss of information</u>. Of course
I can send the ball back up to the edge and make it balance, but not
by using only the information that it has fallen to a specific side,
which is the only aspect of the final state I am interested in for
the decision. If I want to make that lack of interest in anything
else <u>physical</u>, then I arrange for a hole on each side of the edge
and introduce some friction so that the ball will come to rest in
one of the holes. We now have a typical irreversible situation.
That was of course produced by the thermodynamic irreversibility
introduced by the friction. But we don't really need that. This
thermodynamic aspect adds nothing to the essence of the random device. I could just as well use the device without it.

So, a random decision device is unpredictable and irreversible.
It is made to be unpredictable and hence, because it is to make a
definite decision from among a finite number of choices, it has to
be irreversible. Thus, <u>I cannot break the causal chain in one
direction without breaking it in the other</u>. If I introduce a device
that prevents prediction, then it also prevents going back in time
and reconstructing the earlier state.

Let me now return to the temporal order of cause and effect.
There are prople who argue that the alleged experience that causes
always precede effects, is nothing but a firmly ingrained prejudice.
They argue that for any causal explanation given in customary terms
they can substitute an equally satisfactory explanation in which
cause and effect have changed roles, and in which therefore the
cause comes after the effect. I believe that this claim founders
on the issue of simplicity. Feynman-Wheeler electrodynamics is
more complicated that the conventional version in which an accelerated charge <u>emits</u> radiation. Suppose we look at our ESP experiment
again. I make sure by elaborate mechanisms that the throw of the
dice by A is truly random, which in a long enough run may be more
and more accurately verified. My adversary's next move is to reply
that perhaps B writes down his numbers in a random fashion; so how
do I know he is not the cause and it's psychokinesis, after all?
The answer to that gambit is to give A loaded dice. Now the throws
of A will not be random, and if I know the bias of the dice, I can
predict the deviations from randomness of A's throws. My adversary
will have a hard time accounting for these deviations in as simple
a way as I. He has to do it on the basis of some mechanism in B.

I am convinced that this argument, that the temporal order of
cause and effect is essentially a matter of convention, is incorrect.
In my view, as I hope I have convinced you, it is a matter of well
established experimental fact.

Indeed, I may cautiously advance the theory that it is the
order of cause and effect that is at the root of the psychological
"arrow of time." I have never found either of the other two
prominent ideas very convincing: that the direction of time is
determined either by the expansion of the universe, or by the second
law of thermodynamics. The causal order seems to me a much more
plausible explanation. But this is merely a side remark.

Let me, then, return to the special theory of relativity. I will now take it for granted that causes precede their effects, not as a matter of definition but as a matter of extremely well-established experimental fact from which deviations have never been observed. Furthermore, causal chains can be practically, or to any desired degree of approximation, broken by random devices such as roulette wheels. It is these statements, together with relativity, that would be violated by the existence of superluminal signals. The prohibition of such signals does not rest on relativity and logic alone, or on relativity and convention, or on relativity and prejudice, but on relativity and an extremely large body of experience or experimental facts. If such signals were found, we would have to give up either relativity or the conviction that causes always precede their effects, or the belief that we can break causal chains to any desired degree of approximation by suitable random devices and thus <u>experiment</u>. Any of these effects would be quite revolutionary to science.

DISCUSSION

Editor's Note: The following was edited by both the editor and the speaker from tapes of the question and answer period which followed the talk.

DENMAN: The example you use of the ball on the knife edge seems to be a classical analog of a collapsed wavefunction, because there are two final states. It's on one side or the other, and you can't predict which one it is going to go into. You can only look at the statistics afterwards.

A. Well, except that of course, in quantum mechanics you are talking about something that is there <u>essentially</u>, and here in the classical case, you are, in a sense, reducing the information. You say that, of course, there is more information in principle available than on which side of the edge the ball is. In quantum mechanics it is essential and you are at rock bottom, whereas in classical physics you're not at rock bottom, that's not all there is to it.

DRESDEN: They would have a different time evolution. The statement that in quantum mechanics you are indeed in a mixture of states, if you now look at how does this system evolve in time, that is described by an operator on that mixture of states. And that is quite characteristically different from what you would have in a classical situation. Whether it would be one or the other, you might not know, I mean the time evolution operator classically will then act on one of the states and not on the mixture. It's quite different behavior.

DENMAN: The differential equation would act as the time evolution operator and it's the initial conditions, some slight change in the initial conditions....

58

DRESDEN: It is physically different, whether the state is a super-position of two possibilities on which the action must take place, or whether there is one. I mean you can, in fact, construct models in which you can really point out in detail how if you look, say, half an hour later, you can see the physical differences between the one type of time evolution and the other.

A. You have interference in one case and not in the other.

SCHLEGEL: I very much like your statement that expansion of the universe or the law of entropy increase is a correlation rather than a determinant of the direction of time, because it agrees with my own prejudices, but I wondered what further content you could give us in your statement that you thought it was cause and effect that was the primary basis of the sense of time order, time direction.

A. Well, I think that I would simply say that is the way biological mechanisms have evolved. As a matter of fact, effects come after causes. It is useful for living organisms to have their sense of time in that direction. I must say I have not given that any very deep thought exactly what would be the mechanism for producing this sense of time. That would be a very interesting biological question, I must say, but I can't answer it in detail.

NORDTVEDT: Could we put it this way. In the Reichenbach sense, let's say all experience is a whole series of different causal chains. When we try to map points on these chains onto each other, if we couldn't correlate this immense set of chains in an ordered way so that if chain A were mapped onto chain B and chain B were mapped onto chain C, if that didn't give us the same ordering as A on C and we did this for all the chains, I don't think that we would have a sense of time--a flow of time. If all the different causal chains, about half of them ran sort of backwards, with respect to the other when we'd sort of map them onto each other.

A. The world would be different. Essentially what you are saying is that if we lived in a world in which cause and effect relations sometimes go one way, sometimes another way, then presumably we would not have any sense of time at all.

NORDTVEDT: I'm saying the way it comes from mapping one causal chain onto another, you could still have a science in such a world but you would never have evolved the concept of a flow of time.

SAPERSTEIN: You said that cause precedes effect but you put an actual difference in time. Now in the simple $F = ma$, presumable acceleration is the effect and force is the cause at least in the open system. Time is the same. That is a simultaneity. So you are not going to get a directional flow of time from that simple cause and effect.

A. OK, I don't want to prejudice that. I will use precede in the sense of earlier or at the same time, OK?

STAPP: Why didn't you answer Schlegel's question that the sense of time has to do with memory, (I would think.) Memory, if you admit that causes precede effects, naturally goes in one direction. You have the present memory of only things that are caused by what happened earlier. Isn't that the real sense in which cause contributes to time phenomena?

A. Yes, I think that that sounds very plausible. I'm not completely sure that that is not begging the question. But you may well be right.

FADNER: There is another concept in which one can think of time, at least I have, and I'd like to get your comment as to whether you think that's related to your idea that time should be felt as cause and effect. And that is that generally physical systems have a tendency to seek their ground state, as time goes on. Or do you think that's a meaningful statement at all? Do you have any comments on that?

A. Well, that's really again the same type of question as the emission of radiation in electrodynamics and a question of boundary conditions. Of course, I can set up systems, I can set up situations in which that won't be true. I can set up situations in which I close the system, I have a large container and I either produce radiation (electromagnetic field) by boundary conditions on the outside in such a way that it causes a charge in the center to move in a particular way, or I make atoms excite themselves from the ground state by throwing in photons in very specific states, but the point is, of course, that's very, very difficult to set up. They are extremely specific boundary conditions and the information necessary to do it is very, very much more specific than simply the statement that there is radiation there. You have to have specific information on the phases if you do it. If you have sufficiently complicated laboratory mechanisms available, you could do it, but it is very difficult.

FADNER: You see in your explanation you are implying that you can cause something to go up from its ground state.

A. I can, of course I can!

FADNER: So it almost sounds like the "tendency to go down to your ground state" is intimately entwined with the cause-effect relationship that you are proposing.

A. Well, only in the same sense in which the boundary conditions enter in that. I think it is exactly the same question as in electrodynamics.

DRESDEN: What do you have against the description of time in the thermodynamic sense? You said you neither like the astrophysical explanation, of the expanding universe, nor did you like the second law of thermodynamics. Could you explain why you don't?

60

A. I don't see any plausible mechanism, even a faint sign of a
plausible mechanism, that would explain the psychological experience of the direction of the flow of time as connected with the
second law of thermodynamics. Whereas I do see the type of possible connection that Stapp mentioned that has to do with memory and
so on. I could imagine that a mechanism could be established.
Whereas the other I just don't see.

DRESDEN: You are concerned about the psychological aspect of it,
is that correct?

A. Yes, that is the only aspect in which time has an arrow. After
all, in physics there is parametric time, without an arrow. I'm
talking about psychological time. That is, where the arrow of time
really lies. Now you get an arrow of time by definition, of course,
in thermodynamics. Obviously, I'm not talking about that arrow of
time.

HAVAS: I'd like to make three comments which are not entirely connected. Let me make one first. You said something about cause and
effect; you used the Wheeler-Feynman theory as an example, and you
argued about simplicity. It seems to me, that's the type of argument which was used by Mach against statistical mechanics. He said,
there are so much simpler laws which explain the same thing. Now
this, of course, is true up to a point until you get to a situation,
where you say--yes but there are certain things which can only be
explained on the basis of statistical mechanics. I'm unhappy to
use the Wheeler-Feynman theory as an example because there are
various mathematical difficulties in the analysis. But if everything were alright, in what they are doing, then the answer would
simply be yes indeed, looking at it from our point of view, occa-
sionally you would get absorption rather than emission for a particular particle in exactly the same way as you get fluctuations in
statistical mechanics. The other thing, free will, I'm not sure
which way you were actually using this argument but I know that
when I was discussing open and closed systems with other people
very frequently then they say, "Oh yeah", indicating free will, etc.
Now it seems to me that's a complete misunderstanding of what's
happening in physics. We are trying very hard to teach our students
mechanics and thermodynamics to make sure that they understand what
the system is that they are considering. The case that I call an
open system, I mean the experimenter is not part of the system.
Whether there is an overall theory which will finally decide that
whatever he thinks he's doing from free will, is not being done by
free will, that is totally irrelevent in that context. As far as
the consideration of that open system is concerned, all of physics
is based on the assumption that they can do something arbitrarily.
It has nothing to do with the philosophical question of free will.
The third thing was on the theory of time; now that's really a
conference in itself and there have been many. All I can say at
this point is that I would really recommend to people to read
Professor Mehlberg's analyses of the quantum theory of time, which
are very extensive and very much to the point.

A.   To go backwards through your three points: It really has
nothing to do with this conference--and I just threw that out as a
side remark--I realize that one could have lots of discussion about
that and I am also not really that much familiar with everything
that has been said on it. On the question of free will, I absolutely agree with you, of course. That's why I want to avoid ever re-
lying on the feeling of free will. That is exactly the point of
introducing other ways of exerting control with random devices and
so on, or remoteness of mechanisms--rely on that to break any kind
of causal connection. Certainly you would not want to rely on
arguments of free will and certainly that is not ever relevent,
really to any physical discussions. As far as the Feynman-Wheeler
theory is concerned, of course, any argument of simplicity can always be argued about, and people might always disagree with that.
It still seems to me that a conventional type of electrodynamics is
simpler. If there are specific observable differences that is
something else. So long as there are no observable consequences, I
stick with the simpler theory and people may disagree with what's
simpler. Some people may think another theory is simpler, of course.

HAVAS: Let me just clarify one point, of course, when you say that
we take a charge and we move it, then it emits radiation, then you
are absolutely right.   But that's precisely an open system and
Wheeler and Feynman have nothing to say about it.

A.   Sure, sure. In a closed system, of course, then the issue of
cause and effect completely evaporates or becomes completely syme-
tric and you can't distinguish any longer.

ROLNICK: I'd like to discuss or give you a different viewpoint on
the if-and-only-if, the necessary and sufficient connection between
two events. How would you react to this? If A and B are connected
by if-and-only-if, you threw this out (as being a causal connection)
because it was symmetric and you couldn't decide which was the cause
and which was the effect. But if A and B are a time-like interval,
and if I have a concept, because our ideas of cause and effect have
to depend to some extent on our theories of what's the causal connection, if my idea of a causal connection is a disturbance running
from A to B, then temporal order would determine which is the effect and which is the cause for a time-like separation. For a
space-like separation, on the other hand since temporal order is
not Lorentz invariant you might argue against that, except that
supposing I could create a mechanism: If I imagine that there is a
beam, it would have to be a superluminal beam for space-like separation, if I imagine there is such a beam and I put a very thin
screen in the center so I could see whether the beam is on or off
in some way and then I put a thick shield in front of A and the
beam is still there at the screen, then I would say B is the cause
of that beam.   Subsequently if I remove that shield then B would
be the cause of the event at A. And if in fact, the beam goes off
then I put the shield on the other side and I see which way I can
turn off the beam and that is the cause of the beam. Now how would
you react to that viewpoint?

A. Well I think this is really no different. I think that you are
not talking about if and only if in that case; it seems to me that
that really doesn't introduce anything new. As far as the first
part you were talking about, if you feel happy about using a definition like this--in other words, what you are saying is that if you
had an if-and-only-if connection between time-like events and you
like to call the earlier one the cause and the later one the effect,
feel free to do so. I don't think it makes any difference. In the
case of "if-and-only-if", it's purely a matter of definition. I
don't think that that will alter anything. I think that you can do
it either way and if you feel happier with interpretations in that
way, you can do that so long as you have never experienced any
causal chains going backwards in time. In other words so long as
you are absolutely confident that all experience has always indicated that causes are earlier than effects, then of course, you are
perfectly free to use that as a definition in some other circumstances until somebody has found some other causal connection.
Once that has been done however, then you may want to give up that
definition.

DENMAN: Suppose you get cancer if and only if you smoke cigarettes.
Which are you going to cut out?

A. I think it is a question of looking at the small causal steps.

DENMAN: That is my next question, as a matter of fact. Would you
define more clearly what you meant by finding a mechanism between--

A. Oh, just the link, the causal link to reduce the causal chain to
smaller steps. That is really what it amounts to.

DENMAN: But each one is itself a causal step and you can ask what
is the mechanism between those.

A. I don't mean to imply by mechanism any necessary connection or
anything like that. I think that finally when the steps are small
enough, it is essentially again a question of constant conjunction
and of course, theory. Are you happy with the reduction to steps
that you have--once you have reduced the causal chain to steps that
you understand which of course means that you've reduced it to something that's describable in your theory. That's what it amounts to,
I think.

GOBLE: I want to ask you a question about your causal cycle. It's
not clear to me that it makes the theory as closed as you seem to
claim. Suppose it just happened that a tachyon came past A, A of
course was destroyed. But the tachyon came past A to B, at C having received a tachyon you sent a signal to D and that destroyed
the thing at A. It seems to me that there's no contradiction in
this happening. What's more, nothing inhibited the rocketship 2
from pushing his button at time C.

A. No, but of course you are completely neglecting the issue of
control. I mean a tachyon happens to come here. That means that
the signal is not sent from A. A signal is sent from A to B if I

control the switch at B from A, and to have a tachyon just happen
to pass is not such a thing.

GOBLE: The control you are worrying about is whether or not you
can push the button at C which will send one at D.

A. No, no. The button at A. I have a device at A, that button
there, that controls the original signal. OK?

GOBLE: Well the question is what's the original signal? You do
know that you can make a device, which you can push a button and it
will receive a signal. Because you know that you can make a device
that will emit a signal and so by Lorentz covariance you can make
one that will receive one, when you push a button. So what I am
suggesting is suppose you imagine your whole configuration, two
rocketships, as a device in which you push the button for and will
receive a signal coming in from a past where you would have emitted
it at A. Then it seems to me that at no point have you limited
your ability to control the situation in the sense that you are
talking about.

A. No, but the control issue enters when the signal is sent. The
control enters at A. When I say I send a signal that means I control something. I have a random device at A that determines
whether signal 1 or signal 2 is sent. And then I arrange it in
such a way that when signal 1 is sent, what comes back to D is
signal 2 and that then determines what's supposed to be sent. I
get into trouble with the control at A. If I control the whole
thing from the outside, that is an entirely different situation.

GOBLE: It's true, it's certainly true that what I need is a coded
signal coming in past A. But you have already admitted to me that
you can build a device that will receive a coded signal. It seems
alarming, but that's what you have to admit if you are going to
make a device which can emit a coded superluminal signal. So it
seems to me, in terms of just establishing an experiment, building
a set of equipment, you have not restricted the openness of your
theory anymore than you have restricted it by saying I have the
ability to send or receive a coded signal.

A. I must say I don't understand the point. The openness comes in,
in the control at A. OK, the whole thing is simply a matter of
setting up the situation without any control, then all you have to
say is this experiment simply won't go. So, it won't go, for reasons that have to do with the boundary conditions. These boundary
conditions are contradictory. Therefore this thing won't go. This
button won't go. And the experiment simply is impossible, because
the boundary conditions are contradictory. That is all there is
to it.

GOBLE: But what I'm saying is that boundary conditions are not contradictory. The experiment will go and what you have done in setting up your equipment is given yourself a set of boundary conditions where this coded signal is coming in past A.

FADNER: I have kind of a tongue and cheek comment. If we adopt the view that at any instant the various quantum mechanical possibilities will take us into one world or another, this is a very possible scheme if we can play this game and if at A we decide to send such a signal, then we are getting ourselves into a short-ended manifold or fold, or whatever we want to call it, in which we can only go A, B, C, D and then we have to come back A, B, C, D, and we are just kind of oscillating there. But if we decide not to send that signal A, then we can just merely proceed along the different fold which is not dead-ended and so if you want to invent all our manifold worlds, then we can play this game.

A. You've raised the question of which tongue is in which cheek.

MEYER: I agree with the way you proposed to apply causality to physical analysis. Could it be assumed that relativity theory is correct in asserting that no motion rate above a limiting finite signal speed can occur? Then, what is to be made of recent reports of observations by astrophysicists of quasars moving apart relatively at rates greater than this limiting speed.

A. I don't believe them.

SAPERSTEIN: Its just perspective.

A. That is another interpretation of course, but taking literally the way some people have advanced those things, I don't believe them. That's all.

ROLNICK: It is my feeling that your definition of causality when there is sufficiency connecting the two events, is as much a definition as my definition of causality when they are necessarily and sufficiently connected.

A. Well, I mean it is a definition that is true. My argument about sufficient and necessary sounds to me as a convincing way of defining the words "cause and effect" the way one conventionally uses the words, OK. Now I may be wrong on that. Actually, there may be counter-examples to this. There may be examples in which one actually can't avoid the use of the words "cause and effect" in a situation where it is necessary and not a sufficient condition. But I think that in your example, all I can say is that if it is an if-and-only-if situation, and you feel happy with the definition this way, I don't see any way of arguing with you. I think that the only reason you want to do it is because you haven't seen any other kinds of connections, presumably, and in that circumstance I would find that perfectly reasonable.

ROLNICK: Well doesn't that same reason apply to your---

A. Perfectly reasonable, yes. No, I don't think that the necessary and sufficient, that has anything to do with the temporal order with our experience with the temporal order, I don't think that has anything to do with it. I think that you prefer that type of definition because of one's experience with the temporal order, ordinarily. I think that is why you prefer it. Well, if you like that, fine.

GENERAL PHYSICAL PRINCIPLES
AND NON-LINEAR GROUP REALIZATIONS*

Max Dresden
Institute for Theoretical Physics
State University of New York at Stony Brook
Stony Brook, New York 11790

ABSTRACT

The basic question raised in this paper is the relationship between group theoretical and physical notions.
In particular the physical significance of the mathematical entities occurring in group-representations is examined in detail. For this reason a brief outline is
presented of the mathematical definitions and status of
non-linear realizations of groups. Special non-linear
realizations of the Lorentz and Poincaré groups are exhibited. The possible physical meaning of these realizations is discussed. It is shown that there is a fundamental interpretation question involved, which indicates that the identical formalism can describe a wide
variety of physical phenomena. The classical (Wigner)
theory of the representations of the Poincaré group,
allows the description of many-particle systems in terms
of elements and operators in tensor products of Hilbert
space. An appropriate adaptation of this procedure
suggests that the utilization of the non-linear realization of the Poincaré group, describes a relativistic
many-particle system in interaction. General requirements such as causality are shown to be compatible with
the formalism of the non-linear realizations.

1.   INTRODUCTION, MOTIVATION AND BACKGROUND

a.   General Comments

The extent to which general physical principles determine
specific physical features is one of the most important and interesting questions in physics. Historically physics proceeded from
the description and interpretation of detailed phenomena toward
schemes of greater and greater generality. In this development
specific characteristics were frequently shown to be manifestations
of general principles. For example, the many selection rules in

---

*The research reported on in this paper was supported in part by
 NSF Grant GP32998X.

atomic physics are consequences of the spherical symmetry of the
Coulomb forces. In turn these general principles became basic
elements in the construction of many theories. Just what these
general principles are depends on the field of physics considered
and the scope of the investigation.

Symmetry considerations, invariance with respect to particular
groups are commonly imposed general principles. In this paper (as
in many studies in high energy and particle physics) invariance with
respect to the Poincaré group (the inhomogeneous Lorentz group) will
be assumed. Another general principle frequently used in the construction of theories is some form of a causality principle; the
subject matter of this conference. As was known before and became
again apparent during this conference, there is no single, mathematically simple, physically totally satisfactory, definition of
the principle of causality. In the axiomatic version of quantum
field theory,[1] causality (often called locality or microcausality)
is introduced by the requirement that field operators $\phi(x)$, shall
commute at space like separations

$$[\phi(x), \phi(y)] = 0 \quad (x-y)^2 < 0 \tag{1}$$

(x and y label space time points). The metric used is $x^2 \equiv g_{\mu\nu} x^\mu x^\nu =$
$(x^\circ)^2 - (\vec{x})^2$. Physically this condition expresses the fact that measurements of field operators at space like separated space-time
points do not interfere with each other. This has something to do
with causality alright, but it is not obvious that the complete
content of the (or a) causality principle is contained in (1).
Furthermore (1) has something to do with relativity and quantum
theory as well. This is a typical situation, the precise form of
a causality principle tends to be theory and formulation dependent,
so that both the formulation and the content of the causality requirements can be discussed only within the framework of specific
theoretical constructs.

The present study is a preliminary to investigate the constraints imposed by causality (and other related) conditions on a
theory which itself is formulated in group theoretical terms. The
invariance principles are of course directly expressed in terms of
groups, and the combination or campatability of the invariance requirements with the causality requirements is the point at issue.
In order to carry out this program it is essential to first understand the relationship between physical and group theoretical
notions. Once this connection is known one can attempt to incorporate additional physical requirements such as causality in the
same scheme and study the combined system. Consequently a considerable part of this paper is devoted to a study of the relation
between physical and group theoretical concepts.

b.   The Wigner-Dirac Suggestion

The first systematic investigation of the relation between
group theoretical and physical quantities was carried out by Wigner[2]
in connection with the unitary representation of the Lorentz group.
In this and more explicitly in a subsequent paper[3] it was established that the Poincaré group has just two Casimir operators $P_\mu P^\mu$ and
$W_\mu W^\mu$. (Here P_μ, $M_{\mu\nu}$ are the four generators of the translations
and the six generators of the homogeneous Lorentz group respective-
ly, while $W_\mu = \frac{1}{2}\varepsilon_{\mu\nu\rho\sigma} M^{\nu\rho} P^\sigma$). The classification of the irreducible
representations of the Poincaré group proceeds via the eigenvalues
of the Casimir operators.

In addition to characterizing the irreducible representations
these eigenvalues have a direct physical interpretation as the mass
and spin, of a particle. Thus the particle characteristics (at least
the mass and the spin) are obtained as characteristics of irreducible representations, providing an illustration of the connection
referred to before. The mathematical existence of just two Casimir
operators, has as a physical consequence the existence of just two
particle attributes. This same mathematical analysis however shows[3]
that in addition to representations of zero or positive eigenvalues
of P^2, and discrete (integer or half integer) eigenvalues of W [4],
there are also representations describing objects of zero mass and
continuous spin. Although logically there is nothing wrong with
that, it still leads to the somewhat uncomfortable conclusion that
a group theoretical analysis of the Poincaré group, which yields
two known characteristics of existing relativistic particles, also
leads to another class of objects which do not appear to exist.
One can of course just assert that this class of objects is not
realized in nature, but it is more in the spirit of the present
approach to look for general principles which would prohibit the
existence of such objects. It is precisely in this connection that
a form of a principle of causality reappears. Several attempts
have been made to demonstrate that the existence of the zero mass
continuous spin representations would lead to conflicts with locality notions.[5] This is a much more satisfactory reason for the non-
existence of the zero mass continuous spin representation than the
mere assertion that somehow they are not realized in nature. The
arguments just outlined suggest that it is a reasonable procedure
to try to identify or relate group theoretical notions with physical concepts. If some of the group features do not seem to play
any role, it is again reasonable to look for physical (or mathematical) principles which would preclude these features from having
physical relevance. There is of course no guarantee (nor is it
necessary) that all the group theoretical concepts have physical
counterparts, but the previous successful experiences, would suggest that a number of important physical concepts can indeed be
associated with group theoretical notions. This general set of
ideas will be referred to as the Dirac-Wigner suggestion.[6,7] The
main application of the Dirac-Wigner suggestion will be to non-
linear realization of groups. The mathematical theory of the non-

linear realization of groups is nowhere as complete as the linear theory. A brief sketch of some results will be given in section 2. The point to emphasize here, is that in harmony with the Dirac-Wigner suggestion either a physical interpretation of the non-linear realizations should be supplied, or it should be shown that they are incompatible with other physical requirements. As already noted in connection with the non-existence of zero mass continuous spin particles, there is no <u>logical</u> necessity to account for the physical non-occurence of mathematically possible structures. It could be that non-linear realizations of groups just do not play a role in physics. This, although perfectly consistent, is still somewhat unsatisfactory, so that it is well worth the effort to either find a physical interpretation of non-linear realizations, or to show that they lead to physically or mathematically unacceptable results.

Actually it will be shown in section 4, that there are strong hints that non-linear realizations have a physical significance in the description of <u>interacting</u> relativistic systems. It was stressed by Bargmann and Wigner[3] that the irreducible representations they analyzed described non-interacting free particles, and they specifically comment on the possibility that reducibile representations might describe interacting systems. The actual non-linear realizations obtained can then be considered as an infinite sum of reducible representations. The formalism which emerges from the non-linear realizations is reminiscent of that of relativistic particle theories.[8,9,10] It is interesting to recall that in this type theory, (where relativistic invariance is built in from the start and maintained throughout) that there are always questions associated with the introduction of additional constraints such as locality, causality, cluster decomposition properties. This shows that the compatibility of these same requirements with non-linear realization must be studied in harmony with the Dirac-Wigner suggestion. This connection will be discussed in more detail in section 4.

c. Zeeman's Theorem

The question of the <u>compatibility</u> or the relationship of invariance considerations and causality (in one form or another) has been mentioned a number of times. Before embarking on interpretations of non-linear group realizations (especially of the Poincaré group) it is well to note a remarkable result due to Zeeman[11] which establishes a direct connection between a causality principle and the Lorentz group. Specifically the theorem proven is that causality implies the Lorentz group. To see this in more detail let x be a space time point, M, the Minkovski space, i.e. the set of points such that $-\infty < x^\mu < +\infty$, $x^\circ = t$ is the time coordinate. Define an order relation "<" between two space time points x<y by the requirements:

$$x^\circ < y^\circ \tag{2a}$$

$$(y-x)^2 > 0 \tag{2b}$$

where $g_{\mu\nu} = \text{diag}(+,-,-,-)$.

It is easy to check the transitivity of the order relation. Equation (2b) shows that (y-x) is time-like while (2a) states that an event described by x occurs earlier in time than y. Physically an event at x can exert an influence on a subsequent event at y, x and y are (or can be) "causally connected". Consider the class of all mappings F which take M→M, which possess a unique inverse F^{-1} and which satisfy

$$F: \quad M \rightarrow M \tag{3a}$$

$$x<y \leftrightarrow Fx<Fy \quad \text{all } x, y, \epsilon M. \tag{3b}$$

These mappings will be called the causal automorphisms of M. Physically this class of mappings preserves the _temporal_ order or equivalently the _causal_ order of the events in the Minkovski space: events which have a time-like separation keep a time-like separation after the mapping, the time order of the events is preserved, events which are causally connected remain causally connected. The precise result proven by Zeeman is that the set of causal automorphisms forms a group, generated by the orthochronous Lorentz group (including parity inversion but excluding time inversion), by space and time translations, and by the dilatations. This is a remarkable result because neither continuity nor linearity were assumed for the transformations F of M→M, while the final result shows that the transformations F are both linear and continuous.

Thus if the full content of causality is assumed to be invariance with respect to transformations which preserve the temporal or causal order, _in Minkovski space_, this result shows that then Poincaré invariance follows. General as this result is, it must be interpreted with some care. It might appear that there could not exist any non-linear, causal Lorentz transformation on a Minkovski space. Actually it will be shown (in section 3) that such non-linear transformations do exist, but defined on a _subset_ of the Minkovski space. In a similar vein Zeeman's theorem might cause apprehension about causality notions in general relativity, where the transformations are always non-linear. However in the formulation of the theorem the mapping F is explicitly defined as one which takes the Minkovski space M into itself; this requires appropriate modifications for general relativity. In spite of these comments, it will be appreciated that Zeeman's theorem furnishes an interesting and surprising connection between causality and Lorentz invariance.

2. INFORMATION ABOUT NON-LINEAR GROUP REALIZATIONS[12]

a. Mathematical, Formal Aspects

A linear representation of a group G, (elements g) associates a linear transformation T(g) in a linear vector space with each group element g, in such a way that

$$T(g_1 g_2) = T(g_1)T(g_2). \tag{4}$$

Somewhat more formally a linear representation of G is a homomorphism of G into the linear transformations of a linear space. The linear space can be of finite or infinite dimensionality. Similarly a non-linear realization of a group G with elements g, is the association of a non-linear transformation of a space X into itself with each group element. It is necessary to assume certain regularity properties of both X, the representation space, and the group G. Most results are obtained for G a topological group, and G and X as locally compact Hausdorff spaces.

A realization of G on X is defined as a mapping of the Cartesian product (G x X) on X:

$$R : G \times X \to X. \tag{5}$$

Equation (5) means that with every pair (g,x) where g is an element of G, and x is an element of the space X, there is associated a point x' of X. More explicitly with $g \in G$ there is associated a mapping $Fg(x)$ of $X \to X$ such that

$$R: x' = F_g(x) \quad g \in G; \ x, x' \in X. \tag{6}$$

To be a realization F must satisfy the conditions

$$F_{g_2}(F_{g_1}(x)) = F_{g_2 g_1}(x) \text{ for all } g_1 g_2 \in G \text{ and } x \in X \tag{7a}$$

$$F_e(x) = x \quad e = \text{unit element of G.} \tag{7b}$$

$$F_g(x) \text{ is continuous in g and x.} \tag{7c}$$

An association $F_g(x)$ of mappings on X with elements of G which satisfy (6) and (7) is a realization of G on X.

In a linear representation the space X is a linear space and the function $F_g(x)$ is a linear function of x.

To construct "all" the non-linear realizations of a group G, one has to specify both the nature of the space X and the character of the mappings F. The variety of spaces X which can be chosen and the great freedom available in picking the mappings, suggests immediately that a general classification theory must be a formidable problem.

One of the simplest situations arises when the group G is a Lie group and the realization space X is an n-dimensional metric space. In that case the functions $F_g(x)$ are analytic and the relations defining a transformation of the metric space $X \to X$ assumes the form of a coordinate transformation

$$x'_i = F_i(g,x) \equiv F_i(g, x_1 \ldots x_n) \quad i = 1 \ldots n. \tag{8}$$

The F_i in (8) are now analytic functions of $x_1 \cdots x_n$ and $g_1 \cdots g_2$. The composition law of the realization (7) transcribes to

$$F_i(g_2, F_1(g_1, x), F_2(g_1, x) \cdots F_n(g_1, x)) = F_i(g_2 g_1, x_1 \cdots x_n). \qquad (9)$$

Even for this simple case there is no complete classification theory. For example, for the case where G is SU(2) and X the sphere $S^{(n)}$, the results are incomplete. The most complete results are available for <u>transitive</u> realizations. In a sense the transitive realizations play a role similar to irreducible representations in the linear theory. A realization is transitive if in addition to (6) and (7) it satisfies the condition, that for <u>any</u> x, x'εX there exists a g ε G, such that

$$x' = F_g(x) \text{ for all } x, x' \epsilon X. \qquad (10)$$

As g runs through G its image (often called the orbit) covers the space X at least once. If G is a locally compact group, it is not in general possible to construct a transitive realization on arbitrary spaces X. However there does exist an exhaustive enumeration of spaces which can carry the non-linear realizations of those groups.[13] It is at this point that a striking difference between linear and non-linear theory shows up. For a linear representation an arbitrary (unitary) representation is unambiguously specified by its irreducible components. A general non-linear representation can be specified as a union of transitive realizations, however, this union is certainly not unique. The difficulty of obtaining an exhaustive and systematic classification of the non-linear realization (even of Lie groups) is directly related to the non-uniqueness of the composition process.

A final significant notion in group realizations is that of equivalence. Consider a group G, and two representation spaces $X^{(1)}$ and $X^{(2)}$; further let $F_g^{(1)}$ be a realization on $X^{(1)}$, $F_g^{(2)}$ is a realization on $X^{(2)}$. The realizations are <u>equivalent</u> (more precisely <u>continuously equivalent</u>) if there exists a homeomorphism ϕ,

$$\phi : X^{(1)} \to X^{(2)} \qquad (11)$$

such that

$$\phi F_g^{(1)}(x) = F_g^{(2)}(\phi(x)) \text{ for all } g \epsilon G \qquad (12)$$

$$\text{for all } x \epsilon X.$$

Equation (12) asserts that the following diagram is commutative

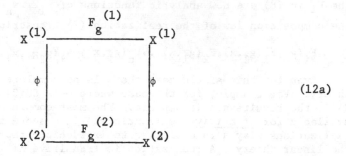

$$(12a)$$

The precise content of the equivalence notion, depends very much on the requirements imposed on ϕ. It is the _existence_ of ϕ which defines the equivalence. If ϕ is required to be differentiable, the appropriate concept is _differentiable_ equivalence. Since continuity does not imply differentiability, realizations can be continuously equivalent but not differentiably equivalent. Equation (12) or (12a) is the generalization to non-linear realizations of the equivalence notion for linear representations. Recall[3] that two linear representations $T_1(g)$ and $T_2(g)$ are called _equivalent_ if there exists a _fixed_ (this means g-independent) unitary operator U, such that

$$UT^{(1)}(g) = T^{(2)}(g)U, \text{ or } T^{(1)}(g) = U^{-1}T^{(2)}(g)U \text{ for all g. (13)}$$

The similarity in the form of (13) and (12) is evident, the mapping ϕ replaces the operator U of the linear case. Distinctions between continuous, and differentiable ϕ have of course _no_ counterpart in the linear case.

b.   Physical Aspects

It is clear even from this brief summary, that a symmetry group defines a number of mathematical entities, some characterize the linear representations, others are features of the non-linear realizations and there might be still others.

Consider a physical system which is invariant with respect to a group G. The fundamental question is whether any physical significance can (or should) be attached to the mathematical notions which accompany the symmetry group. In the linear reqresentation theory, in quantum mechanics, the elements of the representation space themselves have a _direct_ physical interpretation. In the Hilbert space representation the elements are the state vectors of the system. The elements of finite dimensional representation spaces, describe the spin or isotopic spin, or generally the "internal" state of the system. Mathematically speaking, the linear representation space is just an abstract space, the connection with physics is brought in by attributing specific physical characteristics such as quantum numbers to the elements of that space. It is also immediately clear that this method of introducing physical content into the representation space is compatible with linear

realizations only. This must be so, for if the elements of the representation space are identified with physical states, the superposition principle requires that this space is a <u>linear</u> space. Combining this with the invariance properties shows that only linear
representations are possible. It should be stressed, however, that
there exist alternate methods of giving physical significance to
the elements of the realization space. For example xεX, could
<u>label</u> a state, it need not <u>be</u> a state.[14]

It is finally worth noting that the notion of mathematical
equivalence even in the linear case is somewhat tricky. It does
not mean that wavefunctions which are related by a transformation
U (13)   are identical in all respects.[3] This situation becomes
even more delicate in the non-linear case; so that non-linear realizations equivalent in the sense of (12) may well exhibit physically quite distinct behavior. It seems that one can only analyze
this situation by considering specific realizations and specific
interpretations. Examples will be given in the next section.

3. AN EXAMPLE OF A NON-LINEAR REALIZATION OF THE POINCARÉ GROUP

a. Background

In this section some comments will be made about a class of
non-linear realizations of the Poincaré group. A Poincaré transformation will be denoted by (Λ,a). Λ describes a homogeneous
Lorentz transformation; a^μ a translation. Recall that (Λ,a) describes the <u>physical</u> and geometrical relationship between two inertial frames. The group G referred to in the previous section is
the Poincaré group. The space X of the last section will be a four
dimensional Minkovski space B. A vector in the Minkovski space is
called $b^\mu \equiv (b^\circ, b^1) = (b^\circ, \vec{b})$, i=1,2,3. The length of a vector is given
by $b^2 = b_\mu b^\mu = (b^\circ)^2 - \vec{b}^2$. It should be stressed that B is an <u>abstract</u>
space, which has the geometrical character of a Minkovski space;
but it is <u>not</u> necessarily space time. It has been previously shown
that one can construct non-linear coordinate transformations on the
B space, which are non-linear realizations of the Lorentz group.[15]
Actually <u>two</u> results were demonstrated, one for the Lorentz group,
the other for the Poincaré group.

b. The Two Theorems

Let f be a function defined on the Minkovski space B, satisfying the following properties:[16]

1. f is everywhere finite.

2. f has at most a finite number of finite discontinuities.

3. $f(b^\mu) = f(\frac{|\vec{b}|}{b}) = f(\frac{b^i}{b^\circ})$.   (14)

(It is for some purposes useful to assume that f possesses finite
right and left derivatives at $|b^\circ/b| = 1$, but these details will

not be needed in this discussion.)

Let Λ be a homogeneous Lorentz transformation. Then the first theorem asserts that the transformations on the B space defined by (15) constitute a non-linear realization of the homogeneous Lorentz group.

$$b'^{\mu} = (\Lambda b)^{\mu} R(b|\Lambda) \tag{15a}$$

$$R^2(b|\Lambda) = 1 + \frac{1}{b^2} \left\{ f\left(\frac{b^i}{b^o}\right) - f\left(\frac{(\Lambda b)^i}{(\Lambda b)^o}\right) \right\}. \tag{15b}$$

These transformations are not defined over all of the B space; but only over an allowed region α. α is defined by

$$\alpha = \{b: R^2(b|\Lambda) > 0 \ \forall \ \Lambda\}. \tag{16}$$

It can further be shown that the transformations possess an invariant

$$I = b^2 + f\left(\frac{b^i}{b^o}\right). \tag{17}$$

Let (Λ, a) be an inhomogeneous Lorentz transformation (a Poincaré transformation). The second theorem states that the transformations on the B space defined by (18) constitute a non-linear realization of the inhomogeneous Lorentz group.

$$b'^{\mu} = (S^{\mu}_{\nu}b^{\nu}) R(b|S) \tag{18a}$$

$$S^{\mu}_{\nu}b^{\nu} = \Lambda^{\mu}_{\nu}b^{\nu} + a^{\mu} A(b) \tag{18b}$$

$$A(b) = \frac{b^2}{b^2 + f} \tag{18c}$$

$$R^2(b|S) = 1 + \frac{1}{b^2} f\left(\frac{b^i}{b^o}\right) - \frac{1}{(Sb)^2} f\left(\frac{(Sb)^i}{(Sb)^o}\right) \tag{18d}$$

In this case there is an <u>invariant</u> separation between two points b, β of B given by

$$K(b,\beta) = (bA^{-1}(b) - \beta A^{-1}(\beta))^2. \tag{19}$$

The results quoted here exhibit non-linear realizations of the Lorentz and Poincaré groups on subsets of a 4 dimensional Minkovski-space. The space X of the previous section is therefore just the allowed region α of the Minkovski space B. It is possible to analyze these transformations further, for example, the 6 generators $M^{\mu\nu}$ of the Lorentz transformation and the 4 generators P^{μ} of the translations can be constructed as differential operators on the B space:

$$M_{\mu\nu} = M_{\mu\nu}^{(o)} - \frac{1}{2b^2} (M_{\mu\nu}^o f)D \tag{20}$$

$$M_{\mu\nu}^o = b_\nu \frac{\partial}{\partial b^\mu} - b_\mu \frac{\partial}{\partial b^\nu} \tag{20a}$$

$$D = b^i \frac{\partial}{\partial b^i} + b^o \frac{\partial}{\partial b^o} . \tag{20b}$$

The $M_{\mu\nu}^{(o)}$ have the differential form of the usual infinitesimal generators of the linear theory, D is the dilation operator. In (20) $(M_{\mu\nu}^o f)$ means that the differentiation action of $M_{\mu\nu}^o$ includes just f, so that $(M_{\mu\nu}^o f)$ acts as a b-dependent coefficient in front of the differential operator D.

The translation operators assume the form

$$P_\mu = \frac{1}{i} (A(b) \frac{\partial}{\partial b^\mu} + \frac{1}{A^2(b)} \frac{\partial A}{\partial b^\mu} D). \tag{21}$$

The factor $\frac{1}{i}$ is just put in for conventional reasons. A is defined as before (18c). With the explicit form of the generators P^μ and $M^{\mu\nu}$, the various commutation relations can be calculated, and as expected the <u>algebra</u> of the generators (20) and (21) is precisely the usual Poincaré algebra. This confirms that the non-linear transformations (15) and (18), generated by (20) and (21) are a realization of the Poincaré group. Since the <u>algebraic</u> relations between the generators are the same as in the linear case it follows in particular that $P_\mu^2 = P_\mu P^\mu$ and $W_\mu W^\mu$ still commute with all the generators. This result followed just from the commutation rules of the generators.

c.   Remarks

1.   It is worth noting that although the transformations on the B space are badly non-linear the operator form of the infinitesimal generators, shows that in the B space these operators themselves are perfectly good <u>linear</u> differential operators.
2.   It is clear that the realizations constructed have a very special form. Even so the occurence of the arbitrary function f indicates that this is a pretty wide class of realizations. It is worth observing that if f has the special form

$$f(b) = - \frac{(b^o)^2 - \vec{b}^2}{(b^o)^2 + (\vec{b})^2} \tag{22}$$

then the allowed region $\mathcal{a}$ is the exterior of the unit sphere in the B space:

$$\mathcal{a} = \{b^\mu, (b^o)^2 + \vec{b}^2 > 1\} \tag{23}$$

The formal description of the non-linear realizations, including various examples can be worked out along the lines indicated here, but the real important question is the physical interpretation of the formalism.

4. THE INTERPRETATION QUESTIONS. INTERACTIONS

a.   The Physical Interpretation of the B Space

In the last section a non-linear realization of the Poincaré group was exhibited. The realization was in terms of non-linear transformations of the B space. The elements b^μ of the B space were not given any physical meaning. In this section some possible interpretations will be suggested.

1.   A very obvious interpretation is to identify the vectors b^μ of B with the vectors x^μ of a Minkovski space time. It is then understood that $b°=t$, $b'=x'$, so that the elements of the representation space <u>are</u> the result of space-time measurements performed on a <u>single</u> particle. This realization then describes a space-time behavior which is distinct from the usual behavior. The precise behavior depends on the nature of the function f. For the form (22) of f, the result of space-time measurements will always give results outside the unit sphere; one might say that the object described has a finite space-time extension. The description is still a Lorentz invariant description, the invariant is (22).

$$I = (t^2 - \vec{x}^2)\,(1 - \frac{1}{t^2 + \vec{x}^2}).\tag{24}$$

It is particularly noteworthy that the light cone defined by the requirement that I=0, consist of the usual light cone, but it is attached to the unit sphere as (24) shows. It is also clear from (24) that in the allowed region, the exterior of the unit sphere one has

$$\text{sign } I = \text{sign } (t^2 - \vec{x}^2).\tag{25}$$

This means that the separation in time-like and space-like regions is unambiguous in the allowed region but would break down if there could be transformations between the forbidden and the allowed region. This however is impossible, so the causal structure is maintained in the allowed region. Thus one can conclude that if indeed the B space were the physical space-time of a single particle, there would be in principle observable effects. The precise effects would depend on f, and could be quite striking.

2.   Consider the interpretation where the vectors b^μ in the realization space are identified with the energy-momentum four vectors of a single physical particle. Thus $b°=E°$, the observed energy, while $\vec{b} = \vec{p}$, the observed momentum. The non-linear realization now describes the behavior of these physical entities under Lorentz transformations, translations, rotations.

As before this will give rise to observable changes. For example, if f is again assumed to have the form (22), the invariant
of the transformation, now has the form:

$$I = (E^2 - \vec{p}^2)\,(1 - \frac{E^2 - \vec{p}^2}{E^2 + \vec{p}^2}).\tag{26}$$

In this context the invariant must be interpreted as the rest mass
of the particle. Therefore the use of the non-linear realization
predicts a deviation from the usual relation between energy and
momentum. Just what and how large this deviation is depends on the
assumed form of f. In principle experiments could determine what f
is. These two examples should suffice to indicate that the observable effects depend strongly on the physical significance attributed
to the elements of the realization space. In particular experiments
could eliminate whole classes of interpretations and realizations.
Two remarks should elucidate the discussion in this section.

    1. In the theorem of Zeeman the fundamental requirement imposed
was that the automorphisms should be causal, which meant that under
the automorphism F, the relation x<y should be preserved. x<y meant,
$x° < y°$, $(y-x)^2 > 0$. As was noted in the discussion in section 2 these
requirements made a considerable amount of physical sense in terms
of causality concepts. It is clear from Zeeman's proof that if one
uses a space such as B, which has just a <u>geometrical</u> Minkovski character and imposes the condition that under F the relation b<β shall
be preserved, where b<β, means $b° < β°$, $(β-b)^2 > 0$, the result obtained
will be identical. (F is the Lorentz group.) However without a
physical interpretation of the elements of the B space, the preserv-
ation of the relation b<β is really quite arbitrary. Even in the
case that the elements of B are the momenta, the condition to be preserved p<q has no longer a very immediate connection with causality.

    2. In the usual linear theory, there is a well defined connec-
tion between position and momenta. The momenta can be defined either
as the conjugate variables to the position, or equivalently as the
generators of the space time displacements. In the non-linear description these two definitions are no longer equivalent. (A glance
at (21) shows that if $b^\mu = x^\mu$, P^μ is still distinct from $1/i\ \partial/\partial x^\mu$).
So even if the elements b^μ are picked to be the space-time variables,
arguments must still be supplied to decide whether P_μ or $1/i\ \partial/\partial x^\mu$
is the physically measured momentum.

b. Many Particle Systems, Interactions

    An irreducible representation of the inhomogeneous Lorentz
group is characterized by the eigenvalues of $P^2 = P_\mu P^\mu$, and $W^2 = W_\mu W^\mu$.
These eigenvalues are m^2 and $m^2 s(s+1)$ respectively. The generators
can be realized as linear operators in a Hilbert space, whose ele-
ments are functions of p $\phi(p)$, where $p_\mu p^\mu = m^2$. The p^μ are the eigenvalues of P^μ. Since the components of the momentum P_μ commute
there are such simultaneous eigenstates. The Hilbert space is thus
characterized by m,s the mass and spin of a particle. The elements

of the space $H_{m,s}^{(1)}$ are the (relativistic) one particle wave functions. The scalar product is defined by

$$<\phi|\phi> = \int_{p^2=m^2} \frac{d^3p}{p^\circ} |\phi(p)|^2. \tag{27}$$

The integral is over both sheets of the hyperboloid

$$p^2=m^2 \qquad (p^\circ)^2=m^2+\vec{p}^2. \tag{27a}$$

The Hilbert space contains eigenstates of P_μ

$$P_\mu|\phi(p)> = p_\mu|\phi(p)> , \qquad p^2=m^2. \tag{28}$$

Consider next a system of two particles, the Hilbert space then becomes a direct product of the two one particle Hilbert spaces

$$H = H_{m_1 s_1}^{(1)} \otimes H_{m_2 s_2}^{(2)}. \tag{29}$$

The separate Hilbert space contains as elements $\phi_{m_1 s_1}(p_1)$ and $\phi_{m_2 s_2}(p_2)$; vectors in $H_{m_1 s_1}^{(1)}$ are eigenvectors of $(P_1)^2$ and W_1^2, similarly for $H_{m_2 s_2}^{(2)}$. The scalar product in H is

$$<\psi(p_1 p_2)|\phi(p_1 p_2)> = \iint d^4 p_1 d^4 p_2 \delta(p_1^2-m_1^2)\delta(p_2^2-m_2^2) \cdot$$

$$\theta(p_1^\circ)\theta(p_2^\circ)\psi^*(p_1 p_2)\phi(p_1 p_2). \tag{30}$$

The operator of the total momentum in H is

$$P_\mu = P_\mu^{(1)} \otimes 1 + 1 \otimes P_\mu^{(2)}. \tag{31}$$

$P_\mu^{(1)}$, $P_\mu^{(2)}$ are the momentum operators in the one-particle Hilbert spaces (1) and (2). From (31) one has

$$P^2 = (P_o^{(1)}+P_o^{(2)})^2 - (\vec{P}_1+\vec{P}_2)^2. \tag{31a}$$

The problem to be looked at is the relation between the eigenvalues of P^2 and those of $(P_1)^2$ and $(P_2)^2$, in other words the relation between the energies of the two separate one particle systems and the combined two particle systems. For this purpose, let $\phi(p_1)$ be an eigenstate in $H^{(1)}$ of $P_\mu^{(1)}$, let $\phi(p_2)$ in $H^{(2)}$ be an eigenstate of $P_\mu^{(2)}$. Recall that $p_1^2 = m_1^2$, $p_2^2 = m_2^2$. Further assume that there exists a $\phi(p_1 p_2)$ in H (defined by (29)) such that

$$P_\mu^{(1)} \; \phi(p_1 p_2) = p_{1\mu} \phi(p_1 p_2) \qquad (32a)$$

$$P_\mu^{(2)} \; \phi(p_1 p_2) = p_{2\mu} \phi(p_1 p_2) \qquad (32b)$$

$$P^2 \; \phi(p_1 p_2) = M^2 \; \phi(p_1 p_2). \qquad (32c)$$

(One can write $\phi(p_1 p_2)$ as a product $\phi(p_1)\phi(p_2)$ then (32a) and (32b) are immediately satisfied.) Combining (32) with (31a) gives directly (if $m_1 = m_2 = m$)

$$M^2 = 2m^2 + 2\sqrt{(m^2 + p_1^{\,2})(m^2 + p_2^{\,2})} - 2\vec{p}_1 \cdot \vec{p}_2 \qquad (33)$$

In the center of mass system of the two particles $\vec{p}_1 = -\vec{p}_2 = \vec{p}$ and one obtains the well known result that

$$M^2 = 4(m^2 + \vec{p}^{\,2}). \qquad (34)$$

Thus the eigenvalue of P^2, contains both rest energy $4m^2$ and kinetic energy. For $\vec{p} = 0$ at rest there is the obvious result that $M = 2m$.

The reason for repeating these well-known results of the linear representation theory, is to see what if any alterations occur if non-linear realizations are employed. As already stated before, everything that has to do with the algebraic structure will of course remain unchanged as will be the definitions (31). However the elements of the <u>one</u> particle Hilbert space will be functions $\phi(p)$ which are defined not on the hyperboloid $p^\mu p_\mu = m^2$ as before but instead on the surface S

$$p^2 + f\left(\frac{p^\mu}{p^\circ}\right) = m^2 \qquad (35)$$

(m the mass is still fixed).

The scalar product which replaces (27) is

$$\langle \psi(p) | \phi(p) \rangle = \int d^4 p \, \delta_S \theta(p^\circ) \psi^*(p) \phi(p). \qquad (36)$$

Here δ_S is the surface δ function on S. Presumably[17] the Hilbert space $H^{(1)}$ contains a set of basis functions which satisfy

$$P_\mu | \phi(p) \rangle = p_\mu | \phi(p) \rangle. \qquad (37)$$

P_μ is now defined by (21). Consider now a system, composed of two particles. One can now for this two particle system construct the two particle Hilbert space and initiate the procedure outlined before. The basic assumption is that there shall exist states $\psi(p_1 p_2)$ in H, satisfying

$$P_\mu^{(1)} \psi(p_1 p_2) = P_{\mu_1} \psi(p_1 p_2) \quad p_1^2 + f_1 = m_1^2 \tag{38a}$$

$$P_\mu^{(2)} \psi(p_1 p_2) = P_{\mu_2} \psi(p_1 p_2) \quad p_2^2 + f_2 = m_2^2 \tag{38b}$$

$$P^2 \psi(p_1 p_2) = M^2 \psi(p_1 p_2). \tag{38c}$$

From this assumption and the previous definitions one obtains the counterpart of (33) as

$$M^2 = m_1^2 + m_2^2 - f_1 - f_2 + 2\sqrt{(m_1^2 + \vec{p}_1^2 - f_1)(m_2^2 + \vec{p}_2^2 - f_2)} - 2\vec{p}_1 \cdot \vec{p}_2$$

$$f_1 \equiv f(\frac{p_1}{E_1}) \quad f_2 \equiv f(\frac{p_2}{E_2}). \tag{39}$$

If $\vec{p}_1 = \vec{p}_2 = 0$, i.e. the particles are at rest one obtains for $m_1 = m_2$

$$M^2 = 4m^2 - 4f(0). \tag{40}$$

This indicates that f functions as a binding energy, the mass of the two particle system at rest is no longer the algebraic sum. If one picks $m_1 = m_2$ and $\vec{p}_1 = -\vec{p}_2 = \vec{p}$, the analogue of (34) becomes

$$M^2 = 4(m^2 + \vec{p}^2) - 4f(\frac{p}{E}). \tag{41}$$

Again the interpretations of f as an interaction energy is very suggestive.

If all the assumptions made are indeed correct, it would appear that the arbitrary functions in the non-linear realizations have some connection with interactions.

REFERENCES

1. R.F. Streater and A.S. Wightman; P.C.T., spin and statistics and all that, W.A. Benjamin Inc., New York, 1964.
2. E.P. Wigner, Annals of Mathematics, Vol. 40, No. 1, 1939.
3. V. Bargmann and E.P. Wigner, Proc. Nat. Ac. of Science, 34, pp. 211-213, 1946.
4. Actually it is the operator $P^{-1}W$ which represents the square of the spin angular momentum. Its eigenvalues are s(s+1) with s=0,1/2,1.
5. Chakrabarti A., J. Math. Phys. 12, 1813, 1822, 1972; Gallardo, Kalnay, and Risenberg, Phys. Rev. 158, 1484, 1967.
6. Although the implementation of this program was initiated by Wigner, the general philosophy underlying it seems to go back to Dirac. (See the acknowledgement in Ref. 4 and the suggestion that groups of transformations are of special importance in Ref. 7.)

7. Dirac P.A.M., Proc. Royal Soc. of Edingburgh 49, 122, 1938.
8. Foldy L., Phys. Rev. 122, 2751, 1961.
9. Ekstein H., Comm. Math. Phys. 1, 6, 1965.
10. Coester F., Helv. Phys. Acta 38, 7, 1965.
11. Zeeman E.C., Journ. of Math. Phys. 5, 490, 1964.
12. A large number of papers have been written using different aspects of the non-linear representation theory of internal symmetry groups. Since these papers deal with specific applications, they do not contain extensive expository material. The brief summary given here might supplement the discussion in these references: J. Schwinger, Phys. Letters 24B, 473, 1967; S. Weinberg, Phys. Rev. 166, 1568, 1968; Coleman, Wess, Zumino, Phys. Rev. 177, 2739, 1969; Salam and Strathdee, Phys. Rev. 184, 1750, 1969; Khristov and Stoyanov, Inst. Th. Phys. Kiev, I.T.P. 68/56; Michelsson and Wiederle, Comm. Math. Phys. 1970.
13. S. Helgason, "Differential geometry and symmetric spaces", Academic press, N.Y. and London, 1962.
14. It should be noted that although the elements of the realization space are called x, this does not mean at all that these entities are positional variables of any kind.
15. Proofs of these statements can be found in: M. Dresden and A. Albano, Journ. Math. Phys. 13, 275, 1972, and M. Dresden, "Fundamental Interactions at high energy" Coral Gables, 1970 (Gordon and Breach, N.Y. 1970). In these papers the space time character of the realization space was unduly and unnecessarily emphasized. This limited the possible interpretations, so that it is important to stress the abstract "uncommitted" nature of the realization space.
16. The conditions on f are **sufficient** for the discussion, but they are not **necessary**.
17. The analysis which follows makes a number of mathematically unproven assumptions, they mainly have to do with the unproven existence of certain objects.

DISCUSSION

Editor's Note: The following was edited by both the editor and the speaker from tapes of the question and answer period which followed the talk.

HAVAS: I would like to make a comment on Mr. Zeeman. It's been a couple of years since I read it, so I don't remember exactly what the assumptions were, but I think that they are not quite as harmless as meets the eye. I know that people in general relativity have worried about causal space-times a great deal and they come up also with very reasonable definitions which I also don't remember precisely, but for example, there is a paper by Samuels and Kronemeyer, mathematically very difficult, but which comes out with fully causal space-times according to very reasonable definitions of causality in gravitational space. It is not quite as straightforward, as they are reduced....

A. That, of course, may be so but those are the words he says. I don't know if those guarantee that that's all he uses. He does use that it is the whole space. He does use the particular topology of the space.

HAVAS: He does have that all points of the space have the same property.

BELINFANTE: We know of oodles of spaces that have completely different topologies, in particular if there's matter in there and if you have crazy kinds of singularities then your topologies are just as crazy.

A. The topology is indeed something he did not say, but I looked at the proof and he specifically does use (it). However, let me make one comment about this: again the remarks which I made about the spaces B, P, and X; they have something to say also here because (it's amusing) I gave you the definition of causality which Mr. Zeeman gave and I phrased it in terms of the language of space-time but who tells me that's space-time? I can again do it in terms of momentum space, however, in that case the physical causality principle is replaced by the positivity of the mass. That's a much less clear transparent physical picture than the picture which we have right up here (in space-time). It's in fact, in general, very useful. But I think you are right, the topology is certainly explicitly used.

MEYER: Although I claim to be a solid state physicist, my knowledge of group theory itself is actually quite limited.

A. You are not alone in this.

MEYER: The question I wish to ask is for my education and information rather than yours. Analysis was the preferred mathematical procedure for physicists when they took for granted that matter and energy are continuous. My impression is that group theory became more popular when physicists began to adopt the assumption, on good grounds, that actually matter and energy are discrete. What I wonder is, would there be any possibility of getting rid of these undesirable continuous operators and bringing these Casimir operators into sway, if the assumption that we today adopt concerning the structure of space-time, that it is continuous, were given up.

A. I can answer your question most precisely. Of course, I don't know. You see physicists like all other people, are very great at admonitions. We should not be doing this, or more precisely, you should not be doing this. Since I really have enormous trouble understanding most things, except when I deal with moderately controllable and complete physical situations, and I must now admit fully what I said, that philosophy is infinitely harder than mathematics, where at least I know where I am. (I may not be able to do it.) Perhaps, the most ambitious attempt that I know of on this was done by David Finkelstein. David Finkelstein, of course, is an extraordinarily imaginative and brilliant fellow, and he says he never again wants to deal with anything that's continuous. Well, so far this has the main result--he's out of physics. He tries to discover but in a sense, I think, one must do something, you have to start somewhere. To really have such a totally radical change....

SAPERSTEIN: You say that if A occupies all of B, then you can prove that f is a constant or zero. That means that if the allowed space is the entire space, you must have a linear realization. You cannot have a nonlinear realization.

A. Right. And in fact that is in harmony with what, in fact, is in all the books, where one, in fact, proves.... Effectively, the linearity there comes from a translation argument, and for translational arguments all the linear translations get rid of these also. Because you can translate all along when you need. That's right. That's not difficult to do. By the way, I only gave you these. I would not want to create the impression that these are the only nonlinear realizations. No, there are many, many, many others. But you get similar difficulties; they usually have objects in front, so I just mentioned these here.

SAPERSTEIN: This is a pretty general nonlinear one since you haven't specified f.

A. But it's a function. You can, in fact, do much better than this, and I did not want to. If you really want to put this in more group theoretic terms, this always goes if you have a Lie group and if you take the direct product of the Lie group and a one parameter group then you can always get all this. What you're going to have here is that, in fact, the reason this is general is this f is the general invariant function of the dilatation group. You see if you take a function of the ratios, you get a more general thing. I just have given you this one here. Well it is a correct statement that if indeed A is all of B then you prove that f is zero, not one, not a constant, but indeed zero.

BELINFANTE: I have never looked at these pictures that people draw when they draw these collapsing spaces. You see these growth lines and you see the light cones are tangential to the time axis. You don't think that anything like this is....

A. I do think it's connected except I do not exactly know what the connection is. You see, this is really after all very suggestive; I have a circle, right, or a sphere which protects the origin. You cannot get in. You see now this is very suggestive to somehow think that it is not unrelated to something in general relativity. As a matter of fact, these nonlinear realizations are very reminiscent in many ways of matters in general relativity, but I should warn you, that you must (here) very sharply distinguish between local and global considerations: Locally, where the metric is nonsingular (you can also have a metric here) essentially you would have a Riemann-Christoffel tensor which indeed is zero. It blows up, in fact always, at the singular places, and it leads to local and global considerations which are really very different. I think there is some connection except I don't know the connection.

SUDARSHAN: What is the objection that you had about continuous spin? Do you mean continuous spin-spacelike or continuous spin-- zero mass?

A. Continuous spin--zero mass.

SUDARSHAN: Why don't you like it? Second law of thermodynamics?

A. No, I have nothing against them unless Wigner was right. Wigner and Jauch have been writing many, many papers and as I said they proved that continuous spin is incompatable with causality notions, two years ago.

SUDARSHAN: I don't see. Unless you say it offends the black body radiation law or something like that.

A. It doesn't do any such thing. It is perfectly alright. I don't rely on the fact of thermodynamics; that's what you have in mind. No, I guess it was a matter of elegance more than anything else. If you want to admit them then it's certainly fine with me. I, in a sense, mainly used it to set the tone of the kind of argument.

SHPIZ: Here when you combine the two particles, it's very attractive for thinking in terms of interactions. I don't know what your m means if you no longer have the full Poincaré group. You don't have a nonlinear realization of the Poincaré group.
A. Oh yes I do, I most suredly do, P^2 is the mass! You see what we do....

SHPIZ: Would you have translation then and get out of your circle? How do you stay in the circle?

A. You don't. That's a whole subject in itself. But you can ask the question: How are you going to get the Poincaré group out, because I have not talked about it, that you somehow can.

Now the Poincaré group is the following: let me write down here
(Λ, a), that would be a typical Poincaré transformation. Now,
again I make the assertion that you can have a nonlinear realization of the Poincaré group. The points in the circle, you are
never able to see. In fact, let me go back one step. You might
say 'this cannot be!' How can this be Poincaré invariant because,
after all, there is a light cone, right. You have a light cone,
and everybody knows that I can put the light cone wherever I wish.
But now let me try to interpret for you precisely what Poincaré
invariance or invariance under translation tells you. It tells
you, you are at liberty to pick the apex of the light cone anywhere you wish. However, once you've picked it, no longer are
all points in the space, which is now parameterized by your light
cone being somewhere, no longer are all points of the same type.
For example, if I say 'is this point here space-like or time-like?'
that doesn't mean anything. It's only if I say: 'here is the
light cone,' now I can answer your question, you see. When you
talk about the light cone, you must really say it's my light cone!
I can tell you 'get out of my light cone'. (It's not all that
easy to do.) And what Poincaré invariance means--you pick the
origin somewhere and now we may also pick it somewhere else and
there shall now exist an element of the group which will trans-
form the space from this as origin to that as origin. Now I shall
do exactly the same thing with B space. By the way you see its a
funny topology, but I'm now instructed to say I don't like this
point as origin here, I would like this point here. You tell me
where you want it! Then I will construct a translation. It's
not just a translation because also you squeeze things. It's a
translation which in fact doesn't take this point directly here,
because the interior of the protective sphere is never transformed
at all; it will take this space and it will squeeze it together
and you will automatically get this protective sphere around the
new origin. This, by the way also means (a rather delicate point)
that these transformations lack continuity at a certain point;
I'll give you an example: Suppose I take 2 points which are close
together; and I take 2 sequences. And suppose the first sequence
converges to the first point from the left and the second sequence
converges to the second point from the right. Perfectly fine! I
choose to make the first point the origin of my coordinate system.
I told you I can do that anywhere. Now, it is not difficult to
show that this first sequence of points that converged to this
first point, now converges to this point on the protective sphere;
it's the closest that it can get. This second sequence converges
right up here to this diametrically opposite point of the sphere,
and you see in fact what happened. You now no longer have continuity. These two things which were as close as I said before will
now no longer be close then. That's the price you pay, you see.
But you have to prove that. I did not write down the formula.
There's a lot more, but I need that obviously, otherwise I cannot
really talk about anything.

MACROCAUSALITY AND ITS ROLE IN PHYSICAL THEORIES*

Henry P. Stapp
Lawrence Berkeley Laboratory, University of California
Berkeley, California 94720

ABSTRACT

The physical meaning of the macrocausality property of scattering transition probabilities is described, and the role of this property in S-matrix theory and other physical theories is discussed. The macroscopic causality properties of theories with shadow particles, are examined and are shown to contradict the general interpretational principles of quantum theory. Shadow particles have been introduced to remedy the unitarity difficulties of indefinite-metric field theories.

INTRODUCTION

Experience has causal properties, and these should be reflected in physical theory. However, one cannot simply deduce general theoretical causality properties directly from experiment, for experiments are neither infinitely precise nor infinitely extensive. Experiment can merely suggest possibilities, and rule out others.

The form that a theoretical causality property takes will depend on the theoretical structure in which it is imbedded. In fact, a given theoretical structure often suggests a natural causality property. For example, in quantum field theory the natural causality property is that fields at space-like-separated points commute:

$$A(x)\, A(y) \;=\; A(y)\, A(x) \qquad \text{for} \qquad (x - y)^2 < 0. \tag{1}$$

This commutator causality requirement appears to lead to mathematical inconsistencies, and the suggestion is often made that it may be too stringent. For it imposes precise conditions at infinitely small distances, and hence goes far beyond what experience tells us.

This lack of close connection between the commutator causality property and experiment is due in part to the lack of any close connection between the field operators of quantum field theory and experimental observables. This latter deficiency is an objectionable feature of quantum field theory. For a basic precept of quantum theory, at least at the nonrelativistic level, where the mathematical inconsistencies do not arise, is that the basic operators of the theory correspond directly to experimental observables. The logical structure of quantum theory and its connection to experience was built on this premise.

---

*This work was supported by the U.S. Atomic Energy Commission.

To bring relativistic quantum theory into accord with this precept Heisenberg devised S-matrix theory. This theory conforms to the basic precepts of both quantum theory and relativity theory, and it does not encounter the mathematical difficulties associated with the commutator form of the causality condition.

S-matrix theory has no observables corresponding to space-time points or to sharply defined space-time regions. Thus it might seem that S-matrix theory would have no natural causality property. This is not the case: S-matrix theory has a natural causality property, called macrocausality, which in fact plays an important role in the logical and mathematical structure of the theory.

In this talk I shall first describe the physical content of the macrocausality property. This property blends a certain intuitive idea of causality with a specific dynamical assumption. Then I shall discuss the role of macrocausality in S-matrix theory and other physical theories. Finally, I shall apply these considerations to the problem of causality in theories with shadow states.

My subject is narrower and more technical than those of most of the earlier talks, and my presentation is aimed partly at physicists who wish to understand the S-matrix causality concept. However, I shall discuss here only the physical ideas, not the mathematical details,[1] and thus hope to reach also those in the audience whose interests are mainly philosophical. Philosophers should find it useful to have a clear understanding of a causality property that is more elaborate than certain traditional ones, and to see how this causality property is actually used in contemporary physical theory.

MACROCAUSALITY

A. General Remarks

Macrocausality deals only with those observables that occur in S-matrix theory, namely with scattering transition probabilities. These quantities can be measured to high accuracy by means of experimental arrangements of a kind that physicists actually can and do set up. This does not mean, however, that macrocausality can be derived from experiment. For macrocausality is a general property, whereas tests cover only special cases. Moreover, macrocausality refers to asymptotic distances whereas only finite distances are experimentally accessible.

Macrocausality cannot be derived from microcausality. These two causality properties are complementary. Macrocausality deals with arbitrarily large distances, whereas microcausality deals with infinitely small distances. Moreover, as will be discussed, macrocausality is equivalent to a set of analytic properties in the physical region itself, whereas microcausality implies analytic properties only outside the physical region. Thus neither one implies the other. Macrocausality formalizes a certain physical idea, which is discussed next.

B. The Physical Idea

The physical idea of macrocausality is that interactions are transmitted over <u>macroscopic</u> distances only by physical objects. This idea is a macroscopic version of the primitive idea that the world consists only of physical objects, and that these objects act on each other only by direct contact. Two examples will illustrate the main points.

Example I. A baseball is hit into a window. In this example we can identify the following features:
   (a) Cause: The baseball is hit.
   (b) Effect: The window breaks.
   (c) Link: The baseball travels from the bat to the window. That is, a physical object travels from the space-time region of the cause to the space-time region of the effect. This is illustrated in Fig. 1.

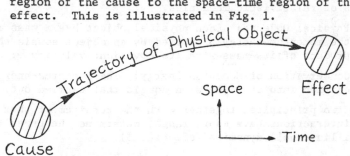

Fig. 1. A physical object travels from the space-time of the cause to the space-time region of the effect.

Example II. A set of billiards balls move about under the influence of their mutual collisions. In this case physical objects travel between the space-time collision regions. This is illustrated in Fig. 2.

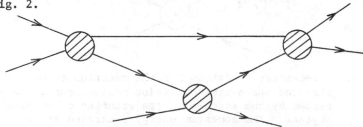

Fig. 2. Physical objects travel between the space-time collision regions. Each space-time trajectory represents the path of the center-of-mass of a physical object.

In these examples a distinction is drawn between long-range interactions and short-range interactions. The long-range interactions are those that are transmitted from one space-time region to a far-away space-time region by a physical object. These interactions fall off (in a statistical sense) at large distances only

by the geometric factor associated with beam spreading. The remaining interactions are those associated with the exchanges of
momentum-energy that occur when the physical objects collide.
These latter interactions are associated, in various theoretical
models, with potentials, or virtual-particle exchange, or unstable-
particle exchange, or nonlocal interactions, or even with a breakdown of the concept of space-time at small distances.

The physical idea of macrocausality is that these remaining interactions are short range. That is, the longest-range interactions
are those carried by physical objects, and hence all interactions
not carried by physical objects fall off faster at large separation
than those carried by physical objects.

To make this idea well defined one must identify the interactions carried by physical objects. This is done by invoking two
basic ideas of relativistic mechanics.

(a) Physical Objects: Each physical object has a mass m, and
the momentum-energy p carried by an object equals the
product of its mass with its covariant velocity v: p=mv.

(b) Conservation of Momentum-Energy: The momentum-energy
carried into any collision equals that carried out.

These two principles, together with the requirement that the
remaining interactions have short range, determine the gross features of billiard ball dynamics (see Fig. 3).

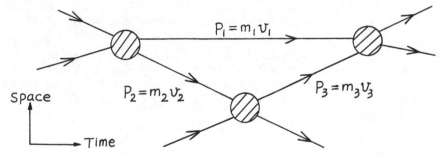

Fig. 3. A necessary condition for the reaction to occur is
that the space-time collision regions can be connected by the space-time trajectories of physical
objects. The momentum-energy p carried by each
object must be directed along its space-time velocity
v, and the proportionality factor must be the mass
of that object. Momentum-energy must be conserved
at each collision.

The finer details of the dynamics will depend on the precise
form of the short-range interactions. However, uncertainties associated with short-range interactions can be effectively damped out
by moving the physical objects farther apart.

This idea can be made precise by considering a set of scattering processes that are related to each other by space-time dilation.
This dilation of the physics can be described by introducing a
"scaled" coordinate system. The scaled coordinates x' are defined
by x = x'τ, where x represents the physical space-time coordinate,
and τ is a scale parameter that tends to infinity. If one fixes the
space-time trajectories in x' space then the physical objects corresponding to these trajectories are moved apart as τ tends to
infinity, unless the trajectories intersect.

Any finite distance Δx shrinks to a point in x' space, as
τ → ∞. Hence the x'-space image of any (finite-radius) physical
object shrinks to a point. And the x'-space image of any finite-
radius interaction region shrinks to a point. Thus if all interactions not carried by physical objects had finite radius then the
necessary conditions for a reaction with specified initial and
final trajectories in x' space to occur for arbitrarily large τ
would be this: the trajectories of the initial and final particles
would have to coincide with the initial and final trajectories of
a "causal network". These networks are defined in and below Fig. 4.

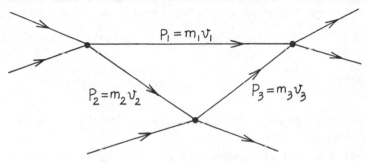

Fig. 4.   A typical causal network. A causal network repre-
          sents the necessary condition for a classical re-
          action to occur if the physical objects are point
          particles that interact only via point interactions.

A causal network depicts the space-time flow of conserved
momentum-energy from initial particles to final particles via a
network of intermediate physical particles. The momentum-energy
p_j carried by each particle is related to its space-time velocity
v_j by $p_j = m_j v_j$. Momentum-energy is transferred between particles
only at points where their trajectories intersect.

The assumption that all interactions not carried by physical
objects have a finite radius is unrealistic and unnecessary. However, some assumption about the way in which those interactions
fall off is needed to give precise content to the macrocausality
property.

The dynamical assumption is now introduced. It is assumed that
all interactions not carried by physical objects fall off at least
exponentially under space-time dilation. This dynamical assumption
is analogous to the assumption that the potentials of non-relativ-

istic theory have Yukawa-type tails.

From this exponential fall-off property one can derive analyticity properties. Weaker fall-off properties yield weaker conclusions. For example, power-law fall-off properties yield continuity properties. However, in what follows the exponential fall off is assumed.

So far the discussion has been purely classical. To pave the way to quantum theory it is useful to exhibit the classical form of the scattering transition probability formula. To do this each initial and final particle j is replaced by a statistical ensemble. This ensemble is represented by a classical probability function $w_j(p,x) \equiv w_j(\vec{p},p^o,\vec{x},t)$, defined by

$$\int_{\substack{\Delta\vec{p} \\ \Delta\vec{x}}} \frac{d^3\vec{p}\, d^3\vec{x}}{(2\pi)^3}\, w_j(p,x) = \begin{array}{l} \text{The probability that a particle from} \\ \text{the ensemble corresponding to particle} \\ \text{j satisfies } (\vec{x},\vec{p}) \,\epsilon\, (\Delta\vec{x},\, \Delta\vec{p}) \text{ at time} \\ \text{t.} \end{array}$$

The particles in these ensembles are free. Thus the energy p^o is fixed by the mass-shell constraint. Moreover, the values of $w(p,x)$ at any one time t determines its value for all times.

The classical transition probability formula is then

$$P[\{w_j(p_j,x_j)\}] = \int \left[\prod_j \frac{d^3\vec{p}_j\, d^3\vec{x}_j}{(2\pi)^3}\, w_j(p_j,x_j) \right] S[\{p_j,x_j\}]$$

where S is the transition probability kernel. In this formula the times t_j can be chosen arbitrarily.

The quantum mechanical transition probability formula can be cast into classical form.[2] The function $w(p,x)$ is defined by a relativistic generalization of Wigner's formula:

$$w(p,x) \equiv \int \psi^*(Mv - \tfrac{1}{2}q)\, \psi(Mv + \tfrac{1}{2}q)\, e^{-iqx}(M/m)^{1/2}\, 2\pi\, \delta(q\cdot v)\, \frac{d^4q}{(2\pi)^4}$$

where

$$M = (m^2 - \tfrac{1}{4}q^2)^{1/2}$$

and

$$v = p/m.$$

The function $S[\{p_j,x_j\}]$ is defined in a similar way:

$$S[\{p_j,x_j\}] = \int \prod_{j=1}^{\hat{n}} \left[\frac{d^4q_j}{(2\pi)^4}\, 2\pi\, \delta(q_j\cdot v_j)\, e^{-iq_j\cdot x_j}\, (M_j/m_j)^{1/2} \right]$$

$$\times\, S(\{M_j v_j \mp \tfrac{1}{2} q\})\, S^*(\{M_j v_j \pm \tfrac{1}{2} q\}).$$

Here the upper sign is to be used for initial particle variables, and the lower sign is to be used for final particle variables, and $S(\{p_j\})$ is the usual S matrix.

C.   Quantum Formulation

The physical idea of macrocausality is expressed in terms of the concepts of classical physics. From this idea one can derive some very general properties of the classical scattering transition probabilities. The quantum theoretical macrocausality property is the statement that these general properties, which follow directly from the (classical) physical idea of macrocausality, are enjoyed by the scattering transition probabilities of quantum theory.

These general properties are of the following kind: they assert that under specified conditions on the initial and final wave functions of the scattering process the scattering transition probability falls off at least exponentially as $\tau \to \infty$, due to the assumed exponential fall off of all interactions that are not carried by physical objects. For under the specified conditions the scattering process can occur only if there is at least one transfer of momentum-energy that cannot be carried by any physical object, yet must carry over a distance that increases linearly with τ. Under these conditions the exponential fall off of the scattering transition probability follows directly from the physical idea of macrocausality.

These considerations can be made quantitative by considering semi-classical models. In these models one allows momentum-energy to be transferred between particles by various possible mechanisms (see Fig. 5).

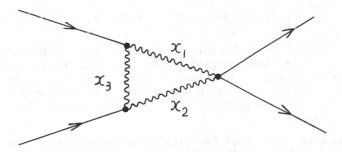

Fig. 5.   Momentum-energy can be transferred between physical
particles (solid lines) by various possible mechanisms (wiggly lines).

The probability that momentum-energy is transferred by a given mechanism is allowed to depend on the momenta p_j of the various particles involved, and on the various space-time intervals x_i over which the transfers carry. However, in accordance with the physical idea of macrocausality, this probability $P(p_j, x_i)$ is required to have a bound that decreases exponentially under space-time dilation:

$$P(p_j, x'_i \tau) \le B(p_j, x_i) \, e^{-\gamma \tau} \; .$$

Here $B(p_j, x_i)$ is bounded in any bounded region in (p_j, x_i) space.

Different mechanisms can have different B and γ, but it is assumed that an upper bound on the scattering transition probability can be obtained by considering, in any finite momentum-energy range, only a finite number of different mechanisms.

Properties of scattering transition probabilities that hold in every model of the kind just described are regarded as general properties that follow directly from the physical idea of macrocausality.

It may be remarked that Planck's constant enters into S-matrix theory only as the parameter that fixes the scale of physical space-time relative to the mathematical space-time variable that occurs in the representation $\exp(ipx)$ of the translation operator. Thus the space-time dilation generated by the transformation $\tau \to \infty$ is equivalent to the transformation $h \to 0$. This means that the macroscopic limit $\tau \to \infty$ is equivalent to a classical limit $h \to 0$. Consequently, the macrocausality property can be regarded as a form of correspondence principle: it asserts that the classical physical idea of macrocausality becomes valid in the classical limit.

III. APPLICATIONS

A. Derivation of Analyticity Properties

To derive analyticity properties from the macrocausality property one uses, for the initial and final particles, wave functions of the Omnes type:

$$\psi_j(p_j, \tau) \equiv \chi_j(p_j) \, \exp(i \, p_j \, a_j \, \tau) \, \exp -(\vec{p}_j - \vec{P}_j)^2 \, \gamma_i \, \tau.$$

The factor $\chi_j(p_j)$ is an infinitely differentiable function that is zero outside some finite region. The second factor generates a space-time translation by the amount $a_j \tau$. These translations move the particles apart in x space, but leave them unmoved in x' space. The third factor is a gaussian which concentrates the function near $\vec{p}_j = \vec{P}_j$ for large τ.

These Omnes functions have important properties. The width in momentum space shrinks like $\tau^{-1/2}$. Thus the width in coordinate space expands like $\tau^{1/2}$. Therefore the width in x' space shrinks like $\tau^{-1/2}$, and the x'-space trajectory region (i.e., the region where the particle is likely to be found) shrinks to a line, as indicated in Fig. 6.

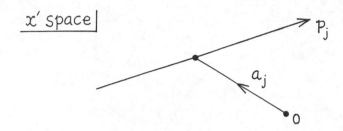

Fig. 6. The trajectory region shrinks to a classical
trajectory in x' space.

More quantitatively, one finds that the probability that the
particle lies in any closed bounded region in x' space that does
not intersect the classical trajectory drops exponentially to zero.
Similarly, the probability that the particle has its momentum p in any
closed bounded interval that does not intersect $\vec{P}$ goes exponentially
to zero. Thus when viewed in x' space the particle goes over, in
effect, to a free particle, modulo effects that fall off exponentially as $\tau \to \infty$.

Using these properties of the Omnes wave functions one may show
that the macrocausality property implies the normal analytic structure.[1] This normal analytic structure consists of two properties.
The first is that the physical-region singularities of scattering
functions are confined to Landau surfaces. These surfaces, discovered by Landau, contain all perturbation theory physical-region
singularities. That is, the functions represented by Feynman diagrams have physical-region singularities only on these surfaces.

Landau derived equations that defined these surfaces. Later
Coleman and Norton pointed out that Landau's equations are just the
condition that the Feynman diagram be interpretable as a causal network. This connection between causal networks and Landau surfaces
is the root of the connection between macrocausality and the normal
analytic structure.

The second part of the normal analytic structure consists of
the iε rules. The rules assert that the physical scattering functions on different sides of the Landau singularity surfaces are all
parts of one single analytic function. Also, these rules specify
precisely how this function should be continued around each Landau
surface to reach the physical scattering function on the other side
of that surface (see Figs. 7 and 8).

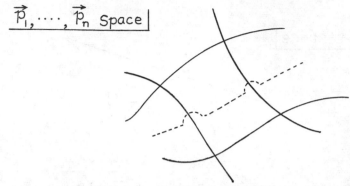

Fig. 7. The iε rules specify the path of continuation that
         connects the physical scattering functions on dif-
         ferent sides of Landau surfaces.

Fig. 8. In an appropriate energy variable the Landau surface
         is (locally) a point, and the physical continuation
         passes into the upper-half plane.

The fact that the scattering function is one single analytic
function is neither trivial nor obvious. In fact, in theories with
shadow particles of the kind discussed in the preceding talk by
Professor Sudarshan the scattering function is not a single analytic
function. Thus these theories lack the macrocausality property. I
shall discuss the consequences of this conflict with macrocausality
toward the end of my talk.

From the normal analytic structure plus unitarity one may derive all physical-region discontinuities. These discontinuities
are the differences between the two different continuations of the
scattering function around a physical-region singularity (see Fig.
9).

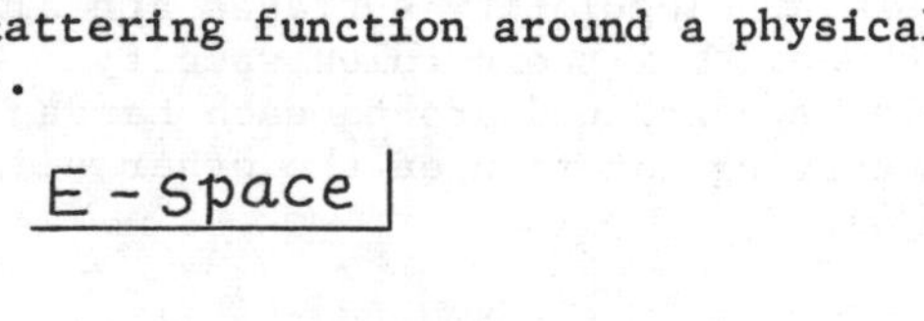

Fig. 9. The discontinuity is
the difference between the
functions obtained by continuing the scattering function
in the two possible ways a-
round a singularity.

Cutkosky obtained formulas for such discontinuities from perturbation theory. However, his formulas were not well defined, and
his arguments were inadequate. Also, they depended on the validity
of perturbation theory. Since these discontinuity formulas play a
basic role in S-matrix theory—discontinuities are the S-matrix
analogs of the potentials of nonrelativistic theory—it is important,
in regard to the matter of internal cohesion, that formulas for them
should be derivable from S-matrix principles.

The simplest and most important discontinuity formula is known
as the pole-factorization theorem. Its simplest case is represented
in Fig. 10.

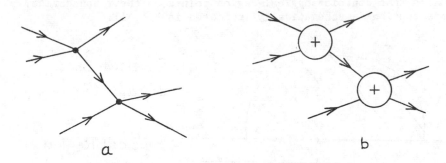

a   b

Fig. 10. The simplest case of the pole-factorization theorem.
Figures a and b represent, respectively, the Landau
diagram (or causal network) and the expression for the
corresponding discontinuity. This discontinuity is
simply the product of the two corresponding scattering
amplitudes, integrated over the intermediate-particle
momentum.

B. Causality Properties of Different Theories

Macrocausality not only implies the normal analytic structure,
it is also implied by it. This means that one can check the causal
properties of a proposed theory by examining its physical-region
analyticity properties: If the theory has the normal analytic
structure then it has the macrocausality property. But if the
theory has the macrocausality property then all long-range interactions are carried by physical particles. Thus the theory possesses all the general causal features that it needs to conform to ordinary macroscopic experience about causality. Any further causality requirement places conditions on the short-range structure of
the theory, and hence extends causality ideas derived from macroscopic experience into realms where empirical support may be lacking.

It is interesting to compare the physical consequences of macrocausality and microcausality. This can be done by considering first
the analytic properties implied by these two causality properties.

The analyticity properties implied by macrocausality are very different from those implied by microcausality. Macrocausality gives analyticity only at physical points (and hence of course in finite, but perhaps very small, neighborhoods of these real points) whereas microcausality gives analyticity only away from the physical points. By counter example it can be shown that microcausality (plus spectral conditions) can never yield analyticity in the physical region itself. Indeed, the primitive domain of analyticity in field theory includes no mass-shell points at all, either inside the physical region or outside it. However, this primitive domain can be extended by methods of analytic completion into mass-shell domains that contain physical-region points on their boundaries. The situation is schematically indicated in Fig. 11.

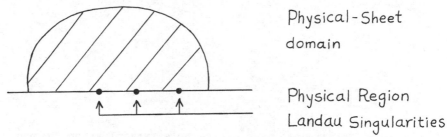

Fig. 11. Macrocausality gives analyticity in the (real) physical region, except at Landau singularities. Microcausality gives analyticity in some physical-sheet domain that contains physical-region points on its boundary.

C. Effects of Poles

To gain understanding of the physical significance of these different domains of analyticity it is useful to consider the effect on scattering transition probabilities of poles that lie in the different regions. Consider, for example, a 2 → 2 scattering process. Suppose, first, that the pole lies at the point $E = m - i\Gamma/2$ in the center-of-mass energy variable. Also, suppose this point is situated on the "unphysical sheet" reached by passing from the physical sheet through the physical region, as indicated in Fig. 12.

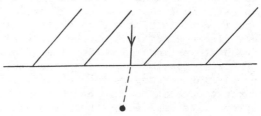

Fig. 12. A pole located at $E = m - i\Gamma/2$ on the unphysical sheet.

Suppose now that the two incoming beams intersect in a space-
time region A, and that the two outgoing beams intersect in a space-
time region B. (The outgoing beams are defined by the acceptance
conditions of the devices that detect the outgoing particles.)
Suppose A and B are both centered around the origin of space (not
time) in some average center-of-mass frame, and that B is later than
A by some average time t, as shown in Fig. 13.

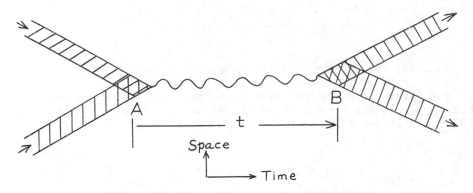

Fig. 13. The two incoming beams intersect at A, and the two
outgoing beams intersect at B. The region B is later
than A by the time t.

If the center-of-mass energy of the pair of incoming particles
is centered around m, and the center-of-mass energy of the pair of
outgoing particles is also centered around m, and if there are no
other nearby singularities, then the scattering transition probability will have the behavior expected from the production and subsequent decay of an unstable particle of lifetime $1/\Gamma$. In particular, for positive t the transition probability will fall off like
$\exp -\Gamma|t|$. (Omnes-type wave functions are used, with $t = \tau$.) For
negative t, on the other hand, the fall off will be much faster,
provided there are no other nearby singularities (on the scale of Γ).
[The rate of fall off is determined by the nearness of the other
singularities, and by the width of the gaussians in the Omnes-type
wave functions.[3]]
Suppose, however, that the pole is situated at $E = m + i\Gamma/2$ in
the physical sheet, as shown in Fig. 14. Then the situation is reversed: for large __negative__ times t the scattering transition pro-
bability will have a term that falls off like $e^{-\Gamma|t|}$, whereas for
large positive times t it will fall off much faster. Thus in this
case the scattering transition probability has the behavior that
would correspond, not to an ordinary decaying particle, but rather
to a particle that propagates __backward__ in time with a decay factor

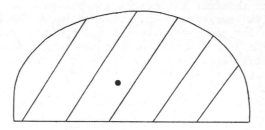

Fig. 14. Pole at m + iΓ/2 in the physical-sheet domain.

$e^{-\Gamma|t|}$. Figure 15 shows the space-time configuration of the incoming and outgoing beams that would reveal this acausal effect of the pole at m + iΓ/2.

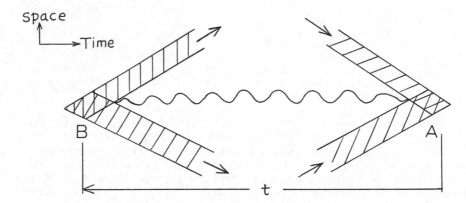

Fig. 15. The effect of the pole at m + iΓ/2. The scattering transition probability falls off like exp −Γ|t| for negative times, where negative times correspond to the outgoing particles being produced before the incoming particles have come together.

Microcausality allows the singularity at m − iΓ/2, which produces the causal behavior, but it forbids the singularity at m + iΓ/2.

If Γ is sufficiently small, and hence the lifetime 1/Γ is sufficiently long, then the acausal effects of this singularity should, in general, be observable. However, if Γ is large then these acausal effects would be hard to observe.

For a 2 → 2 reaction the pole cannot lie right in the physical region itself because of stability requirements. But if two external particles are added, in the manner shown in Fig. 16, then the

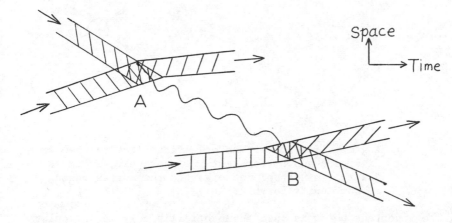

Fig. 16. Generalization of Fig. 13.

intermediate particle pole can lie in the physical region (i.e.,
$\Gamma \to 0$). In this case the exponential decrease factor turns into
the geometrical factor corresponding to the classical spreading of
the intermediate particle beams. In particular, a pole at $m - i\varepsilon$
(where ε is infinitesimal) has an effect on the $3 \to 3$ scattering
transition probability of precisely the kind that would be caused
by a classical particle of mass m being produced at A and absorbed
at B.

D. Theory of Measurement

These physical-region singularities at points $m - i\varepsilon$, and the
pole-factorization theorem expressions for their discontinuities,
play a crucial role in the theory of measurements. Bohr and
Heisenberg emphasized that the consistency of quantum theory requires that the boundary between the quantum system and the (classically treated) world in which the quantum system is imbedded can
in certain circumstances be shifted, so that what was originally
part of the classically treated measuring device becomes part of
the quantum system under consideration. This requirement was studied by von Neumann, in the framework of non-relativistic quantum
theory.

The S-matrix study of the requirement is based on a generalization of the pole-factorization theorem, a special case of which
is illustrated in Fig. 17.

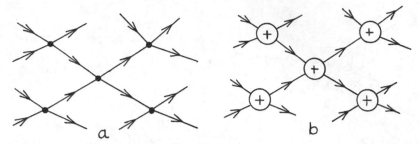

Fig. 17. A generalization of the pole-factorization theorem.
Figures a and b represent, respectively, a Landau
diagram (or causal network) and the corresponding
discontinuity formula.

The important point is that asymptotically only the singular
part of the scattering amplitude contributes. Thus, if the space-
time separations between the five collision regions in Fig. 18 are
all large (see Fig. 18) then the scattering amplitude for the over
all 6 → 6 process can be replaced by its singular part, which is

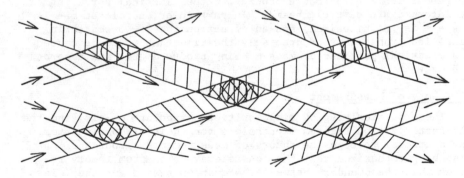

Fig. 18. A space-time process corresponding to Fig. 17.

exhibited in Fig. 17b, and the transition amplitude takes the
factorized form

$$\int \prod_{i=1}^{4} \left[\psi_i^{(*)}(p_i) \, \frac{d^3 p_i}{2p_i^{\,0}(2\pi)^3} \right] \, S(p_i)$$

where

$$\psi_i^{(*)}(p_i) \equiv \int S_i(p_i, \, p_{ij}) \left[\prod_{j=1}^{3} \psi_{ij}^{(*)}(p_j) \, \frac{d^3 p_j}{2p_j^{\,0}(2\pi)^3} \right].$$

Here $S(p_i)$ is the S matrix for the central $2 \to 2$ process in Figs. 17 and 18, where the index i runs over the four outer processes. The four functions $S_i(p_i, p_{ij})$ are the S matrices for these four outer $2 \to 2$ processes. The various $\psi_i^{(*)}$ are ψ_i or ψ_i^* according to whether particle i is an incoming or outgoing particle for the central reaction, and the $\psi_{ij}^{(*)}$ are ψ_{ij} or ψ_{ij}^* according to whether ij labels an incoming or outgoing particle of the <u>i</u>th outer reaction.

The first two outer reactions (reading from left to right in Fig. 18) can be regarded as the reactions in which the two incoming particles of the central reaction are prepared. And the final two outer reactions can be regarded as the reactions that detect the two outgoing particles of the central reaction. Thus the factorized formula for the transition probability shows the consistency between the interpretations in which the outer reactions are considered, alternatively, as integral parts of the overall $6 \to 6$ process, or as the reactions that prepare and detect the incoming and outgoing particles of the central $2 \to 2$ reaction.

IV. THEORIES WITH SHADOW PARTICLES

A. The Measurement Problem

Theories with shadow particles encounter problems concerning measurement, which will now be discussed. The discussion is based on the foregoing discussion of the theory of measurements.

Consider a theory with shadow particles, of the kind discussed by Professor Sudarshan in the preceding talk.[4] Suppose there is a shadow particle of mass m. Consider a $3 \to 3$ scattering process in which the three incoming particles and the three outgoing particles are all ordinary (i.e., non-shadow) particles. And suppose the incoming and outgoing beams are arranged as shown in Fig. 19.

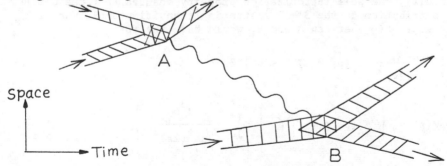

Fig. 19. The incoming and outgoing beams of the $3 \to 3$ scattering process are arranged so that two of the incoming beams and one of the outgoing beams intersect in a space-time region A, and so that the other two outgoing beams and the other incoming beam intersect in a space-time region B. The outgoing beams are defined by the acceptance conditions of the measuring devices that detect the outgoing particles.

Suppose the momentum-energies of the three external particles that intersect at A are such that a particle of mass m and momentum-energy k could be produced in this subreaction. Suppose also that the momentum-energies of the three external particles that intersect at B are such that their momentum-energy imbalance would be corrected by an extra incoming particle of mass m and momentum-energy k. Furthermore, suppose the locations of A and B are such that a space-time trajectory with direction v = k/m connects A to B, as shown in Fig. 20.

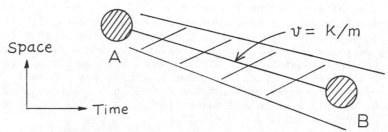

Fig. 20. The space-time region B lies in the region where a particle of mass m and momentum-energy k could go if it were produced in A. The region of space-time corresponding to the various values of k that are compatible with the momentum-energy ranges in the incoming and outgoing wave functions is also shown.

If the particle of mass m were an ordinary (i.e., non-shadow) particle, then there would be pole in the scattering amplitude at m − iε. The effect of such a pole is to give a contribution to this scattering process of exactly the kind that would be expected if a particle of mass m were produced at A and absorbed at B. In particular, the pole-factorization property ensures that the dominant contribution to the 3 → 3 scattering transition amplitude, for large separation between A and B, would have the form[5]

$$\int \frac{d^3p}{2p^0(2\pi)^3} \ \psi^*(p) \ \emptyset(p) \equiv \ < \psi \ | \ \emptyset >$$

where

$$\emptyset(p) = \int S_A(p, \ p_j) \ \prod_{j=1}^{3} \left[\psi_j^{(*)} \ \frac{d^3p_j}{2p_j^0(2\pi)^3} \right]$$

and

$$\psi^*(p) = \int S_B(p, \ p_j) \ \prod_{j=4}^{6} \left[\psi_j^{(*)} \ \frac{d^3p_j}{2p_j^0(2\pi)^3} \right]$$

and all momentum-energy vectors are on-mass-shell. If the formula for $\psi(p)$ is substituted into the expression for the transition amplitude $< \psi \mid \emptyset >$, then the result can be interpreted by saying that a particle of mass m and wave function $\emptyset(p)$ is produced in the reaction at A and detected in the reaction at B.[5]

If the particle of mass m is a shadow particle then the rules set forth by Sudarshan and co-workers[4] say that the S-matrix for this $3 \rightarrow 3$ process should be calculated by using the principle-value resolution of the pole singularity at E = m. That is, one should use

$$\frac{1}{2} \left[\frac{1}{m - i\varepsilon} + \frac{1}{m + i\varepsilon} \right]$$

instead of the usual retarded propagator resolution $[1/(m - i\varepsilon)]$.

The effect of this change on the transition probability rates predicted under the conditions represented in Figs. 19 and 20 is to decrease them by a factor of four. For in these situations only the retarded part of the propagator contributes significantly, and hence the factor of one-half occurring in front of the retarded part of the principal-value propagator produces a factor of one-quarter in the scattering transition probabilities. This means that the shadow particle can be detected by its interaction at B with ordinary particles, but that the probability of its being found at B is decreased by a factor of four.

The fact that the shadow particle can be detected in this way far away from the region in which it was formed conflicts with the ideas of shadow theory. For shadow particles are supposed to contribute to the dynamics, yet not appear as physically observed particles.

The problem, however, is that dynamics cannot be separated from observation. For what is observed is dynamical effects. If the long-range dynamical effects corresponding to a particle are present, then this particle is present. For in quantum theory a physical particle is nothing more than the physical effects that we associate with a particle.

The point, then, is that the effect of the retarded part of the principal-value propagator is to ensure that the shadow particle will propagate through the space-time region indicated in Fig. 20, in the physical sense that it can be detected in this region by probes consisting of ordinary particles.

Since the long-range dynamical effects corresponding to the reaction at B are present it is hard to understand how a charged shadow particle could fail to produce also tracks in a cloud chamber. For the two effects do not seem qualitatively different.

The obvious way out of these difficulties is to make the masses of all shadow particles complex. Then these particles would be unstable, and hence would not contribute to the asymptotic states. This is the strategy of Lee and Wick.[6] But Sudarshan and co-workers do not require their shadow-particle masses to be complex, and in fact usually deal with cases in which the shadow-particle masses are real.

B. Causality Problem

The difficulties just discussed arise from the retarded part of the shadow-particle propagator. The advanced part leads to other difficulties.

The advanced part of the shadow-particle propagator produces acausal precursor effects. In particular, it generates contributions to reactions of the kind shown in Fig. 21.

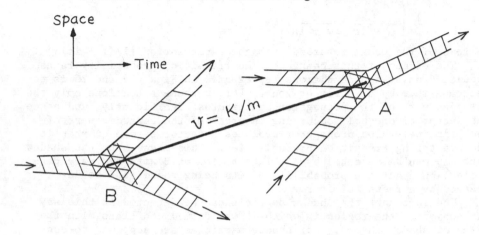

Fig. 21. A scattering process exhibiting the acausal effect. In this 3 → 3 process two of the incoming particles collide at A, and one outgoing particle emerges. The missing energy-momentum k is compatible with that of a shadow particle. The other two outgoing particles are observed to emerge from a region B, which lies in the intersection of the third incoming beam with a space-time trajectory that starts at A and moves backward in time along a space-time line that is parallel to the momentum-energy vector k.

The problem, now, is that the outgoing particles from B can, in principle, be detected before the incoming beams aimed at A are turned on. Moreover, the experiment can be set up so that these incoming beams are turned on if and only if the particles from B are not detected. On the other hand, by making the incoming beams sufficiently intense one can arrange that quantum theory will predict this: if the incoming beams are turned on then particles from B will almost surely be detected, but if the incoming beams are not turned on then particles B will almost surely not be detected.

This gives a "causal loop" similar to those discussed in earlier talks: if particles from B are detected then the beams will not be turned on, and quantum theory will predict that particles from B will almost surely not be detected. Conversely, if particles from B are not detected then the beams will be turned on, and quantum theory will predict that particles from B will almost surely be detected.

It is logically impossible for these statistical predictions of quantum theory to be borne out in a sequence of repetitions of this experiment. Thus quantum theory must, by logical necessity, fail to correspond to experience in the way that quantum principles demand. Thus the introduction of the principle-value propagator in the manner prescribed by shadow theory is incompatible with the basic interpretational principles of quantum theory.

The above argument is based on the Copenhagen interpretation of quantum theory. That is, quantum theory is viewed as fundamentally a procedure by which scientists make predictions about what they will observe under specified conditions, and the wave function is viewed as the quantum theorist's representation of an idealization of the finite system that he is examining, rather than some absolute representation of the world itself.

This Copenhagen view places the scientist and his macroscopic measuring devices outside the quantum system. Thus the quantum system is "open", in the sense used in earlier talks. The scientist sets up the experimental conditions and is, as far as quantum theory is concerned, a free external agent.[7]

The causality problem just discussed, unlike the measurement problem discussed earlier, is not resolved by simply making the shadow-particle masses complex. For if the unstable shadow particles have sufficiently long lifetimes then by making the incoming beam sufficiently intense one could, in principle, still construct experimental arrangements that would lead to the contradictions with quantum theoretical principles. Moreover, even for shadow particles with small lifetimes there are two-particle branch-points at $m + m^*$ = 2 Re m that lie in the physical region itself, and which would give acausal effects that have a power-law fall off, rather than an exponential fall off.[6] Though these acausal effects would in practice be small, they would generally lead in principle to causality problems of the kind just discussed.

A central question to which this conference has addressed itself is whether causality requirements have the force of logical necessity, or are mere expressions of convention or prejudice. Logical necessity can, of course, operate only within a logical or theoretical framework. However, within a given general theoretical framework causality requirements can be a logical necessity. The example discussed in this section illustrates this point.

FOOTNOTES AND REFERENCES

1. The mathematical details are given in D. Iagolnitzer and H.P. Stapp, Commun. Math. Phys. 14, 15 (1969).
2. H.P. Stapp, Foundations of S-Matrix Theory. I. Theory and Measurement, Lawrence Berkeley Laboratory preprint LBL-759, June 13, 1972.
3. These results follow directly from the procedures of Ref. 1.

4. Shadow particles are discussed in literature in the following papers: E.C.G. Sudarshan, Fields and Quanta 2, 175 (1972); C.A. Nelson and E.C.G. Sudarshan, Phys. Rev. D6, 3658 (1972); E.C.G. Sudarshan and C.A. Nelson, Phys. Rev. D6, 3678 (1972); A.M. Gleeson, R.J. Moore, H. Richenberg, and E.C.G. Sudarshan, Phys. Rev. D4, 2242 (1971).

5. These questions are considered in more detail in H.P. Stapp, Phys. Rev. 139, 257 (1965); and in D. Iagolnitzer's article in Lectures in Theoretical Physics, 1968, XID (Gordon and Breach, 1968), ed. K.T. Mahanthappa and W.E. Brittin.

6. T.D. Lee in Proceedings of the International School of Physics "Ettore Majorana", 1970, ed. A. Zichichi (Academic Press, New York, 1971). Cf. also T.D. Lee and G.C. Wick, Phys. Rev. D3, 1046 (1971).

7. There is an opposing naive view of quantum theory that holds that the entire world is represented by a wave function. This view entails, however, either that the superposition principle fails to hold universally, in which case the theory is not quantum theory, or that the world we know is one of a continuously infinite collection of similar worlds, all but one of which must remain forever unobservable.[8] The need to accept such a metaphysical assumption is a big price to pay for shadow particles. Moreover, a technical problem arises. For this interpretation is based on the idea that there is a Schroedinger equation that governs the temporal evolution of the world's wave function. Shadow theory, on the other hand, has been formulated in the S-matrix framework. Thus additional work would be needed to show that shadow theory can be generally formulated in terms of an evolving wave function of the world.

8. These opposing interpretations of quantum theory are discussed in H.P. Stapp, Am. J. Phys. 40, 1098 (1972). References are given there.

DISCUSSION

Editor's Note: The following was edited by both the editor and the speaker from tapes of the question and answer period which followed the talk.

SUDARSHAN: I completely disagree with all the things that you said about the shadow theories. First: This shows that one is not able to make a small change in one part of the theory without, in a sense, a thorough overhaul of all the theory. If you keep the old part of the theory and then make modifications only in one place, then of course you are going to get into difficulties. The physical picture of what actually goes on has to be revised in such a fashion that it is consistent with everything. I completely agree with you that if you make the corresponding analysis that you have made, you do get into trouble. One asks the question --

what happens when you have a shadow propagator which is being cut?
There is a technical correction. It is not just that one line
which has principal value. It is the totality of all the lines
which you are cutting, because you have a space time diagram that
you are cutting at one time. The object that you obtain in this
particular process of cutting is precisely the same thing that
you would get if you took the wavefunction that I had written down
in the non-relativistic model and ask the question: what happens
to this when t → −∞. What you get is something which is not conventional; it is not a series of plane waves, but in fact plane
waves plus a certain amount of spherically converging waves. You
may say at this point, I don't like this theory because I don't
like theories in which you happen to have <u>this</u>. This would be
very much like people saying I don't like quantum theory because
p and q cannot be simultaneously determined. This is a different
theory. If you take this interpretation, then of course in cutting
the lines that you happen to have, there is no problem. I have
not analyzed the question with regard to the unstable particle
going backwards and forwards. If you say that that is so, I
suspect that it must be so. I really have not analyzed it. But
with regard to the question of particles making tracks in cloud
chambers and then being called shadows, this does not happen.
You may dislike this interpretation -- in fact, Dr. Newton and I
had talked about it earlier. In this specification of states,
what is to be identified with the initial state of scattering with
a number of electrons in the incident beam on a target? How is it
to be represented? If it is to be represented as a plane wave
plus spherically diverging waves, then obviously this formalism
is not adequate, because it contains some standing waves. It
contains both spherically converging and diverging in it. But
if you do allow this, then the rest can consistently work. I have
one very good excuse for not continuing this discussion. I have
to catch a plane....

A. Well, I feel it is a little unfair for me to stand up here
alone and argue this point. I would rather Professor Sudarshan
were here to give his point of view. But in the descriptions of
shadow theory as contained in his papers, he specifically says
that you can calculate (in fact that's a big point) the S-matrix
by the usual perturbation prescriptions. And if you do this, for
the simplest case where you have just one particle in the intermediate state, then you certainly get the propagator as I've described it. This is the answer to his point that you have to
consider the whole system with the shadow state in it: in this
particular process I'm calculating, the effect of that requirement
is just to put the one-shadow-state propagator in. For only one
propagator occurs when you calculate the S-matrix in this case.
So if you calculate, according to his prescriptions then you do
get both the advanced and retarded effects that you would expect
from the symmetric propagator.

110

The further question is the problem of how is one supposed to use
the theory. In the usual theory we know very well what to do -- a
good theory, basically, should be one in which you know how to
calculate things. The procedure is this: You set up counters in
a certain way, and set up devices which produce the incoming waves,
and then in terms of the things that you actually observe, you
write down the wavefunctions for the particles produced and for the
particles detected. Then you fold these wavefunctions into the
S-matrix and square, to get the transition probability. So if
Professor Sudarshan wants to say that everything has to be changed,
then he has to tell us how he wants to change everything, in order
that one can actually calculate physical observables. None of what
he just said is contained in any of his papers on these questions.
Those papers lead to the impression that one just calculates in the
physical domain in the same way that one usually does, but with a
new S-matrix. What he is now proposing seems different from our
ordinary ideas of how, in terms of what you actually observe, you
write down the appropriate wavefunctions, which you then fold into
a well defined S-matrix in order to calculate the transition probability. If he wants to go to a different conceptual framework,
then he should tell us what that framework is.

NEWTON: In regard to macrocausality, I'm not sure that I really
understand exactly what you are saying. Can you say specifically
what the consequences of your macrocausality are? Let's take
specifically the case of 3-3 scattering. The diagrams--these
causal diagrams--which you drew seem to be like double scattering
diagrams; on-energy-shell double scattering diagrams. First of all
the double scattering kinematics picks out specific energy momentum
relations between the final and initial states. So are you saying
that when those energy momentum relations are satisfied -- that
then the double scattering term in the   3-3   amplitude domin-
ates? That is question number 1. Now question number 2--what
happens at other energy momenta that can't fit double scattering?

A. The answer to your first question is yes, provided the appropriate space-time conditions are also satisfied. Let me draw a
picture of what we are talking about. [See Fig. 22] This is the

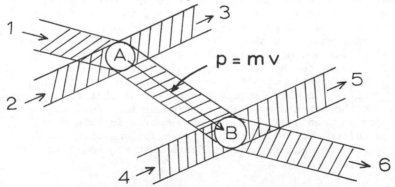

Fig. 22. A space-time picture of a double scattering process.

most primitive form of this macrocausality condition. What we have is two initial wave packets that intersect in a space-time region A. We also have a detector which is accepting the beam labelled 3. Similarly, down below, there is an incoming beam, and there are two outgoing beams. Each of these six beams has a certain momentum spread, which could be fairly narrow. Then, as you say, in regard to the momentum, there are certain conditions. Let us suppose that there is only one diagram that is relevant namely this one-particle-exchange diagram. Then the values that are allowed for the intermediate-particle momentum p are those corresponding to momentum defect of the external particles that interact at A (or at B). If you were to draw corresponding space-time velocities v, defined by v = p/m then these v would define a certain cone of allowed directions. Now if you arrange your final apparatus--by which I really mean the devices associated with the lower part of the reaction--so that region B lies inside this cone of velocity vectors, originating at A, then you expect to see a 1/r fall off. In other words, if you place your detectors so that B lies in the allowed region, then you expect to detect the particle, with probability having a r^{-2} fall-off. On the other hand, if you move region B out of the velocity cone, by moving the detectors, then the transition probability for the whole reaction is expected to fall off faster than r^{-2}. The macrocausality condition is the statement that the (theoretical) transition probability does indeed fall off faster than r^{-2}, and in fact falls off exponentially if you use the Omnes wavefunctions. That is, if you put this region B any place else except in this allowed velocity cone region, you get an exponential fall off, because stable particles cannot transmit the momentum energy p between A and B, and all other interactions are assumed to have an exponential fall off. This is the physical statement of macrocausality. If you put in this assumption, then you get out the analyticity properties I have described. If you put in the analyticity properties I have described, then you get out these fall-off properties.

NEWTON: The only consequence as far as the asymptotic amplitude is concerned is the analyticity property. You are not saying anything about dominance in that amplitude by the double scattering term, or anything like that.

A. That also follows: It comes from the discontinuity formulas. If you apply unitarity in conjunction with these analyticity properties, then you can derive all the physical-region discontinuity formulas. The transition amplitude in the asymptotic domain is controlled by the physical-region singularities, and hence by the discontinuities around these singularities. One can show that if the region B is place inside the appropriate velocity cone then the asymptotic transition amplitude is given by the double scattering term, just as one would expect. However, this result is not put into the macrocausality property. Macrocausality only asserts fast fall off when the causality conditions are not satisfied: it does not, itself, say anything about the cases in which the causal space-time conditions are satisfied.

HAVAS: If I understood you correctly, it's very important for you that in some limit you get agreement with classical predictions, and more implicit in your discussion of that, there seemed to be the idea that you think it is quite clear what the classical predictions are. But I have tried to be very explicit in showing that there are very serious difficulties. Now I don't want to repeat myself, but just to give one example, you talk of an exponential fall off with distance. This is not a statement which has an obvious relativistic meaning. It has a meaning in a preferred frame of reference, for example....

A. May I please interrupt—I said under space-time dilation.

HAVAS: I don't see how this could help. You are talking about four-dimensional distances.

A. Four-dimensional distances, but more precisely, space-time dilations.

HAVAS: If you are talking about four-dimensional distances, that leaves the light cone invariant. You could have a zero distance on the light cone and still have interactions, so it is not really clear to me what this would imply.

A. I mean more than distance. I talk in terms of space-time dilations. You draw your classical diagram, and then I dilate the whole diagram.

HAVAS: I understand that, but it is certainly not clear in relativistic dynamics what the behavior of, let's say, three particles is, with interactions which are functions of the four-dimensional separation.

A. That is correct. We do not know behavior of interactions in relativistic dynamics. Let me emphasize, as I said at the beginning, that I am not claiming in any sense that this macrocausality condition is something that can be derived from experiment. I am saying only that it does convey a certain classical concept of causality. This concept is not necessarily the right one, or the only one. But it does convey a certain idea of what you mean by causality, and this idea does account for everyday cause-and-effect situations. This idea is that reactions are carried over long distances only by stable objects, and the rest falls off faster. To be sure, to derive precise consequences from that general idea, you have to put in also what I term a dynamical assumption. This dynamical assumption is that the remaining things have an exponential fall. Now if you look in various theories you find this is true. I know of no theory in which it's not true. You always find that you have particle effects which have a 1/r fall off, and the rest, in the way I've done things, have the properties that macrocausality asserts: They do fall off with exponential tails. Now just because models have this property it doesn't mean that the world has it. But it is a general property that has a well defined meaning experimentally. You know that if this property is satisfied, then you're not going to have really big acausal effects.

All acausal effects will fall off fast. So macrocausality does
convey a certain idea of causality, and this concept turns out to
be equivalent to an interesting analyticity concept. So I'm not
maintaining that every theory must, by logical necessity, obey
this causality property. But if we impose it then we can pursue
S-matrix theory. I mean, then we have the initial S-matrix analyticity property.
Now, other people would look at the problem differently, other S-
matrix people. They are content to just assume the analyticity
property. If you want to assume the analyticity property, then you
use the theorem the other way, and deduce this causality property.
This property may not be the one that you might want, but it is a
certain causality property. It says that there are not really
big violations of our most primitive idea of causality, which is
the idea that causal effects are carried by stable particles.

DRESDEN: I have two questions--one is that I am a little bit ap-
prehensive about the use of the Wigner function which you wrote
out. If I understood it, you used the Wigner function in order to
provide an appropriate probability density in phase space, in a
quantum-mechanical sense, and then the ideas which you imposed on
S-matrix theory are to conform to that. Is that a correct statement?

A. That's correct. But let me say that I wrote down that particular form in order to be specific. However, the way in which you
make the transition between the classical and quantum probability
functions can be done in a variety of ways. Different people have
done it differently. These variations all have the effects of
merely adding polynomial factors: they merely add extra polynomial factors to the expressions I have used. But these various
forms all lead to the same results, for exponential tails are
essentially unchanged by these polynomial factors. Thus the connection between analyticity and macrocausality is unaffected by
the choice of probability function.

DRESDEN: My question is in part connected with the loss of p-de-
pendence. That is, whenever you take these limiting processes,
either from quantum to classical or from relativistic to nonrelativistic, these limiting processes are always gruesomely nonuniform.
So, you see, if I know the answer, then I know where to go, because I have some feeling of what I mean by classical causality.
I then, essentially, can devise a process that will give what I
want. But I worry about the necessity. My second question is
really the following: It has to do with what you mean by a nice
analyticity notion. Now, I believe that if I just look at what
experiments have to tell me--let me now forget about causality
altogether--then the only analyticity notion I really need is
essentially anything that will guarantee me crossing symmetry. I
have to be able, somehow, to get from one physical region in one
reaction to another. I couldn't care less if I have to go by a
very tortuous path; that is perhaps my tough luck. I couldn't care
less if there are branch cuts and poles and what-not, as long as I

can get from one to the other. That would in principle be enough. Your analytic formulation of course allows that, but it's a great deal stronger than that. Is that correct?

A. The analyticity property that I get, the one that is equivalent to macrocausality, holds only right in the physical region itself. So it doesn't allow you directly to say anything about crossing.

DRESDEN: That analyticity is not really what I want. What is the reason that I want analyticity at all? It's that I believe that various reactions are related by analytic continuation, and I would like to be able to derive this property from causality, and would later like to use this analyticity to derive dispersion relations.

A. Your question is what analyticity do you want. In order to get, say, dispersion relations for two to two amplitudes you need a great deal of analyticity. That analyticity that you can get from microcausality. So, if you want that analyticity, then micro-causality is the thing that you need, or some S-matrix equivalent, which I haven't talked about today. By the way, it wouldn't be sufficient to say only that by some tortuous path you could get to the crossed region. If you want dispersion relations, then you want much more than that. You want the physical sheet to be com-pletely clean, with no singularities except perhaps a few simple ones that you really know about. But in any case, let me return to the question: What analyticity do you want? At least within the S-matrix framework, it seems that one should start in the physical region, and say that this is the region where you have your basic analyticity. Thus one would like to have some reason for expecting some analyticity there, and would like to know where the physical region singularities are. Then one can build on this and say that elsewhere in p-space there are no singularities except those that follow in some sense from the singularities that you already have in the physical region. In other words, you can say that no other singularities will be allowed unless they can be derived as a consequence of those that you already have at the beginning. So, in this sense, you want the analyticity properties in the physical region as your starting point.

QUANTUM THEORY WITH SHADOW STATES: A SEPARATE REALITY

E.C.G. Sudarshan[†]
Center for Particle Theory and Department of Physics
University of Texas, Austin, Texas   78712

ABSTRACT

Starting with an analysis of the natural (but not inevitable!) assumptions of conventional quantum theory the desirability of a generalization is highlighted. I give an exposition of a new scattering theory in which a new class of states enter; these states are relevant for the dynamical description but do not contribute to probability. The transition amplitude in this theory is calculated; it obeys the unitarity relation. Its relation to the standard scattering amplitude enables us to calculate it by simple methods and to study its piecewise analyticity. Certain causality questions are discussed and certain paradoxes resolved. A number of deserving candidates for the role of shadow particles are listed. It is suggested that perhaps "general interpretational principles of quantum theory" are not as general as they should be or could be.

INTRODUCTION: THE NATURE OF DYNAMICAL INTERACTIONS

It is a matter of everyday experience that bodies exchange energy and momentum on collisions. These impulsive interactions have led physicists to postulate action-by-contact as a fundamental law of interaction. The question naturally arises whether all interactions can be brought about in this fashion. Again, it is a matter of everyday experience that there are interactions between objects which do not collide: like the action of gravity or electricity. To bring these under the mantle of action-by-contact we introduce the gravitational or electric "field" which interacts with material bodies by contact. By thus enlarging the set of "physical objects" we can say all these interactions are only by virtue of action-by-contact.[1]

We observe that for these interactions apparently at-a-distance the mediation is by a field. And fields and particles are different kinds of objects in classical physics.

[†]Supported by the U. S. Atomic Energy Commission.

In quantum theory fields and assemblies of identical particles describe the same physical situation; at this stage we may say that all interactions are "by contact" though now we include under the notion of contact creation and destruction of particles. Particle spectra and particle interactions are now no longer independent but are two aspects of the same set of phenomena. We may refer to the observation of particles both directly (by kinematic studies on suitable entities) or indirectly (by seeing their dynamical effects in interactions between other particles). It is then an article of faith, supported by some empirical evidence, that all particles seen indirectly must be seen directly. Attractive and economical is this item of faith: so much so that we usually raise it to the status of a guiding principle.

Despite this temptation there are several reasons to be cautious. Perhaps the most compelling one is that local relativistic field theory inevitably leads to infinities in any honest attempt to calculate the dynamical consequences of any postulated interaction including the quantum electrodynamic interaction. There are ways of extracting useful predictions regarding some physical quantities which are in excellent agreement with experimental results; but nevertheless there are quantities for which the predictions yield infinities. The root of these infinities can be traced to the infinite number of degrees of freedom[3] of fields and to the local coupling which involves all of them.

The kind of troubles one finds may be illustrated by a point particle in interaction with other fields, say the electron in interaction with the electromagnetic field. An honest assessment of the theory suggests that the electron should explode; in other cases instead of an explosion an implosion is predicted. Fortunately electrons and other particles neither explode nor implode! But we must search for a generalization of the local interaction model which incorporates this happy state of affairs. Possibly non-local field theories?

If the principle of deriving all interactions by exchange of particles embodied in local quantum field theory leads to difficulties we can ask: could we include additional forces not derivable from exchange of particles? Can we have genuine non-local interactions?

I have studied non-local theories within the framework of special relativity.[2] Since manifest relativistic invariance for a non-local theory implies both retarded and advanced interactions the conserved quantities of energy, momentum and angular momentum involve non-localized quantities. They can however be rendered local within an extended formalism in which new auxiliary fields are introduced which are coupled to the particles and fields of the theory. Thus, apart from certain additional boundary conditions pertaining to these auxiliary fields that are to be incorporated, the non-local theory becomes an extended local interaction theory.

But we might ask: what about the convergence that arises in theories with suitable non-locality? How does that come about in

the extended local field theory? The answer to this is that the
extended field theory contains "negative weight" fields whose
quantum theory is to be formulated in a linear vector space with an
indefinite scalar product. Such "quantum theories with indefinite
metric" have been studied for a long time,[3] but the resolution of
the paradox of negative probabilities has come only recently.

Even more generally, we can ask the question: is the world of
dynamics of interacting elementary entities composed of (observed)
particles only? Are all shadows cast by entities with substance?

In the following talk Professor Stapp[4] refers to "...the primitive idea that the world consists only of physical objects, and
that these objects act on each other only by direct contact." He
then refers to the generalization in which we allow other interactions to be present provided they fall off with distance like a
power law or an exponential law. He chooses the dynamical assumption: "It is assumed that all interactions not carried by physical
objects fall off at least exponentially under space-time dilation."

But I like to consider the possibility that the world of elementary entities consists of both ordinary physical objects which
are directly seen and non-ordinary objects which can be seen only
indirectly. This "separate reality" simplifies the formulation of
the dynamical picture of the world.

If we take such a world full of shadows with and without substance and insist on a pragmatic world-view with only the ordinary
physical objects the price we pay for it is a non-local interaction
which falls off only according to a power law. This leads to continuous but piece-wise analytic transition amplitudes in quantum
field theory.[2,4]

I have a number of aesthetic reasons to prefer considering the
world to include such nonordinary objects as shadows without substance, to consider the world to be full of mystery.[5,6] I cannot
ask that you share this predilection with me. But I can, and I
will, ask that you agree with me that this possibility is worth in-
vestigating if only for you to reassure yourself that it is not
useful. But I hope that you find it otherwise.

In this paper I aim to show that not all particles "seen" need
be seen directly. This departure from tradition is not to be undertaken lightly, and one should be critical of the prediction of only
piece-wise analyticity. The departure consists in an altered perspective rather than in an unusual interaction structure. The
possibility of such a nonordinary perspective relating to scattering
and transition processes obtains since the determination of transition amplitudes involves not only the natural evolution (the "free
Hamiltonian") but also the boundary conditions. It is in the recognition of this freedom that my theory of shadow states[2,7] has its
foundation; and utilization of this freedom gives form to the theory.
While local quantum field theory with indefinite metric[3] was the
motivation I would like you to see the theory in its generality, to
see the possibility that there could be particles which affect the
dynamics without being seen directly irrespective of whether these
particles have positive or negative norm states.

QUANTUM THEORY OF SCATTERING

Consider a quantum system described by a total Hamiltonian

$$H = H_o + gV.$$

For the time being we need not distinguish between quantum field theories and quantum mechanical systems with a finite number of degrees of freedom except to note that in the former case the various terms would be integrals over all space of corresponding densities. H_o is chosen so that it has the same spectrum as H and with the same multiplicity. [In the field-theoretic case this implies that all mass renormalization terms are included in V.] Let us now choose corresponding (improper) eigenvectors χ, ϕ of H and H_o:

$$H\chi = E\chi \; ,$$

$$H_o\phi = E\phi \; .$$

A formal choice of χ is as a solution of the equation:

$$\chi = \phi + gGV\chi$$

where G is Green's function obeying

$$(E - H_o)G = 1.$$

We verify that this is so by showing that

$$(H - E)\chi = (H - E - gV)\chi + g(E - H_o) \, GV\chi$$

$$= (H_o - E)(\chi - gGV\chi) = (H_o - E)\phi = 0 \; .$$

The "free state" ϕ and the "fully interacting state" χ differ and this difference $\chi - \phi$ may be thought of as a scattered wave:

$$\psi = \chi - \phi = gGV\chi \; .$$

It may be thought of as a free wave generated by a source:

$$\xi = (E - H_o)\psi = V\chi \; .$$

The "scattering" (transition) amplitude from an initial state i to a final state f is then given by

$$T_{fi} = <f|T|i> = (\phi_f, gV\chi_i) \; .$$

This expression can be rewritten in a slightly different form by writing down the formal solution:

$$\chi = \phi + gGV\phi + g^2GVGV\phi + \ldots$$

$$= (1 - gGV)^{-1}\phi .$$

Then

$$\xi = V(1 - gGV)^{-1}\phi = (1 - gVG)^{-1}V\phi$$

and

$$T_{fi} = (\phi_f, V(1 - gGV)^{-1}\phi_i) = \langle f|V(1 - gGV)^{-1}|i\rangle$$

so that

$$T = V(1 - gGV)^{-1}$$

is the matrix of transition amplitudes. All these calculations are standard and they are the generalization of the usual method of taking the coefficients of the asymptotic diverging spherical waves in the usual elementary wave mechanical description of scattering.[8] The scattering is dependent on the interaction gV as well as the free Hamiltonian H_o.

What is important to recognize is that there is still some freedom in defining the correspondence $\chi \to \phi$ and hence of the "source of the scattered wavelets" ξ. This freedom stems from the freedom in the choice of the Green's function G. We may choose any Green's function; all the relationships would continue to hold as long as the defining relation

$$(E - H_o)G = 1$$

is maintained. As G changes so do ξ and T. The scattering is de-fined not only by H_O and H but also the specific choice of G. Since G may be identified with the propagation function, we can interpret this freedom by remarking that scattering depends not only on the interaction but also the manner in which the wavelets propagate[9]: and that is eminently reasonable!

It is natural to ask at this stage as to why this freedom does not seem to show up in the familiar elementary derivation of the scattering of a plane wave by a potential. The solution is unique since we take it for granted that the propagation of the scattered wavelets is forward-in-time and so it is mandatory to choose the so-called retarded Green's function. We choose as the solution χ of the exact Hamiltonian a state tending to a plane wave ϕ in the far past, and to the plane wave ϕ plus diverging spherical waves ψ in the far future. Here the choice is already made and no freedom is left.

The theory of shadow states[2,7] has its origin in the recognition of this freedom; and in the willingness to keep an open mind about whether all particles have to obey this boundary condition.

The transition amplitude matrix T satisfies certain non-linear conditions which incorporate the law of conservation of probability in the elementary theory of scattering. They are therefore called the generalized "unitarity" condition. These relations relate to the difference between T and $T^\dagger$. We have,

$$T^\dagger - T = gV\{(1 - gG^\dagger V)^{-1} - (1 - gGV)^{-1}\}$$

$$= gV(1 - gG^\dagger V)^{-1}(G^\dagger - G)gV(1 - gGV)^{-1} .$$

In writing down this relation we have made use of the hermicity of the interaction gV. The relation could be simplified to read:

$$T^\dagger - T = T^\dagger(G^\dagger - G)T .$$

This non-linear relation depends on the antihermitian part of the Green's function G. When we make a new choice for G we get a new relationship for T. This equation is the generalization of the "optical theorem" which relates the imaginary part of the elastic forward scattering amplitude to the cross section.[8] The optical theorem is directly related to probability conservation. Hence any freedom in the choice of the Green's function and the consequent change in the generalized unitarity relation entail a new probability interpretation. The possibility of discounting certain states from contributing to the probability arises out of this circumstance.

Let us, therefore, pay special attention to the choice of Green's function. The most familiar one is to choose the retarded Green's function:

$$G_R = (E - H_o + i\varepsilon)^{-1} .$$

For this choice

$$T_R^\dagger - T_R = 2\pi i T_R^\dagger T_R .$$

Together with time-reversal invariance this leads to the optical theorem. The probability interpretation is then the usual one with all states contributing to the physical probability. All states are physical states; and the physical probability that is summed over all these states is conserved. We may well say: all virtual states can become real. Or equally well: the world is made up of physical states only.

A NEW SCATTERING THEORY

Now let us be adventurous and explore other possibilities. Let us separate the vector space of states W for world into two subspaces R (for real) and S (for shadow) by means of a projection operator σ:

$$S \quad \{\sigma\psi| \qquad \psi\epsilon \; W\},$$

$$R \quad \{(1 - \sigma)\psi| \; \psi\epsilon \; W\}.$$

Since

$$\sigma^2 = \sigma = \sigma^\dagger \; ,$$

it follows that these two subspaces are orthogonal. We take care
to have σ commute with the free Hamiltonian H_0 so that R and S
are invariant under the free Hamiltonian evolution. In a relativistic theory we may choose σ to commute with the generators of a
larger invariance group including the "free" inhomogeneous Lorentz
group. There is now the possibility of choosing the Green's
function

$$G = \sigma\bar{G} + (1 - \sigma)G_R$$

$$= 1/2 \; \sigma(G_R + G_A) + (1 - \sigma)G_R \; ,$$

where

$$G_A = (E - H_o - i\epsilon)^{-1} \; .$$

Equally well we may write

$$G = (E - H_o + (1 - \sigma)i\epsilon)^{-1} \; .$$

The transition amplitude depends on σ in a non-linear manner and we
must now seek out the proper probability interpretation associated
with the framework.

Before doing this we note one point: the "scattered wave" ψ
for an "incident wave" ϕ is now different from what it would have
been with the standard (retarded) choice for G. This includes some
advanced waves which are expected from the advanced components in
the Green's function we have chosen. Hence the "free" state ϕ and
the state χ in the far past no longer coincide; but the lack of
this coincidence depends on the space S of shadow states. The
correspondence between the "free states" ϕ and the "interacting
states" χ is now more subtle: but it is of course well defined
when we have settled on the projection σ.

We now complete the formulation of the new scattering theory
by specifying the new probability interpretation. We compute the
probability using $(1 - \sigma)$ as the metric operator. The entire subspace of states S now correspond to zero contribution to the probability while the subspace of states R has the usual probability
interpretation. There are no probability amplitudes connecting
R and S. The space S of states, devoid of probability interpretation, is a part of the world that is a mathematical auxiliary.

TRANSITION AMPLITUDES IN SHADOW STATE THEORY

One feature of the theory should be stressed though it should be clear: the Green's function G refers to the entire system and propagates the quantum state of the system. It is not the propagator of a particle, but the propagator corresponding to the system of particles. Misunderstanding on this point seems to prompt many authors including Professor Stapp[4] to criticize shadow state theory incorrectly. The probability interpretation is for the state not for a particle!

This completes the formulation of shadow state theory.

The discussion so far has been in terms of Hamiltonians and interactions and it looks non-covariant. Is the theory in fact, covariant? We may also wonder whether for the new theory we must develop an entirely new computational calculus similar to the one that we have developed, say in perturbation series for a relativistic quantum field theory. The theory is in fact relativistic. To show this and other features of the scattering amplitude in shadow state theory we study the scattering amplitude in perturbation theory carried out in the interaction picture.

Let $gV_I(t)$ be the interaction in the interaction picture. We rewrite the shadow theory Green's function in the form:

$$iG(E) = \int_{-\infty}^{\infty} \{\theta(t') - 1/2\ \sigma\} e^{-i(H_0-E)t'}\ dt'$$

Following the work of Richard[10] we then get the modified Dyson formula:

$$-iT = \sum_{n=1}^{\infty} \frac{(-ig)^n}{2\pi} \int_{-\infty}^{\infty} dt_1 \cdots \int_{-\infty}^{\infty} dt_n\ V_I(t_1)\ \{\theta(t_1-t_2)-1/2\ \sigma\}$$

$$V_I(t_2) \cdots \{\theta(t_{n-1}-t_n) - 1/2\sigma\} V_I(t_n)\ .$$

This can be simplified into the form

$$-iT = \sum_{\nu=1}^{\infty} (-\pi i)^\nu (-i\tau)(\sigma\tau)^\nu$$

where

$$-i\tau = \sum_{n=1}^{\infty} \frac{(-ig)^n}{2\pi} \cdot \frac{1}{n!} \int_{-\infty}^{\infty} dt_1 \cdots \int_{-\infty}^{\infty} dt_n\ T(V_I(t_1), \cdots, V_I(t_n))$$

is the Dyson expression for the standard transition amplitude computed using the fully retarded Green's function (for the system!). It is familiar to us in quantum field theory in terms of Feynman's diagrammatic computational calculus. (We stress that the interaction $gV_I(t)$ in the interaction picture does not depend on the choice of the Green's function.) We thus obtain the simple formula for the physical scattering amplitude:

$$\mathsf{T} = (1-\sigma)T(1-\sigma) = (1-\sigma)\tau(1-\pi i\sigma\tau)^{-1}(1-\sigma)\ .$$

While we have used perturbation theory and interaction representation to calculate T in terms of τ we could derive this relationship in other ways not dependent on perturbation theory. We may take this to be an exact relation.

It is clear that T so defined is relativistically invariant and possesses invariance under internal symmetry operations that were common to σ and V. Though it is not explicit in the appearance of the projection operator, σ is a <u>piecewise constant</u> operator considered as a function of the four-momenta of the particles. Hence the transition amplitude is piecewise analytic since τ itself is an analytic function.

But let us show that T conserves probability. By direct calculation we obtain:

$$\mathsf{T}^\dagger - \mathsf{T} = (1-\sigma)\{(1+\pi i \tau^\dagger \sigma)^{-1}\tau^\dagger - \tau(1-\pi i \sigma \tau)^{-1}\}(1-\sigma)$$

$$= (1-\sigma)(1+\pi i \tau^\dagger \sigma)\{\tau^\dagger - \tau - 2\pi i \tau^\dagger \sigma \tau\}(1-\pi i \sigma \tau)^{-1}(1-\sigma)$$

$$= 2\pi i \, \mathsf{T}^\dagger \, \mathsf{T} \, .$$

In other words <u>the probability</u> with the metric $(1 - \sigma)$ in the space W (or equally well with metric 1 in the space R and 0 in the space S) <u>is conserved</u> by the transition amplitude. And this is equally true whether we are talking about nonrelativistic multichannel systems, relativistic multichannel systems or relativistic field theory. The only word of caution is that in the last case we must ensure that we have a theory which yields a finite amplitude τ.

PIECEWISE ANALYTICITY

The scattering amplitudes in shadow state theories are only piecewise analytic. The conventional amplitude computed from a local relativistic field theory is, apart from phase space factors, an analytic function with singularities which are themselves dictated by physical considerations. In the present theory wherever the shadow states are energetically forbidden the shadow theory amplitude coincides with the conventional amplitude; but for domains where they are energetically allowed the two amplitudes differ thus exhibiting clearly the discontinuity in analyticity. In each piece we have a non-linear function of τ which is therefore analytic.

Since the discontinuity in analyticity enters only through phase space factors the analytic jump starts out with zero. Hence though we have only piecewise analyticity we do have continuity.

Such violations of global analyticity should be experimentally tested. M. G. Gundzik and I have shown that such discontinuities cannot be ruled out even in such well studied cases like pion-nuclear forward scattering.[11] More systematic study of this question would be desirable.

CAUSALITY

The problem of "causality" in such theories is of great interest. The philosophic principle of causes having well defined effects following them, itself, though taken for granted in physics, has been severely criticized by careful and wise thinkers.[12] A discussion of these questions is beyond our scope here: but it does pertain to incompatibility of the twin requirements of logical distinction between the cause and effect and of their invariable association. Seeking for causal connections involves identifying components of a total reality, as <u>autonomous</u> and modifiable, identifying them as causes, and looking for responses to such causes. We have remarked above that the Green's functions we choose are for the whole system (and <u>not</u> for individual particles). If we insisted on incorrectly identifying individual shadow particles as autonomously propagating then one is led to apparent paradoxes. The state has to be treated as a whole. (For states involving shadow particles we do <u>not</u> have a pole factorization theorem of the kind that Professor Stapp[4] presupposes.)

In a previous section we had seen that the piecewise analytic physical scattering amplitude T had only the discontinuities connected with the physical states. The Landau rules would surely indicate a singularity even for shadow intermediate states: but examination of the nonlinear relation between T and the amplitude τ shows that if τ had a shadow Landau singularity the corresponding discontinuity is eliminated from the physical amplitude τ.

The moral of the story is that we can use the geometric picture of colliding billiard balls for the globally analytic mathematical amplitude τ but one must not get misled into making incorrect physical deductions from the elegant geometrical picture. Shadow state theory does not admit the same geometrical pictures permitted by the old theory. There are other entities apart from physical particles that transmit interaction. Incidentally, <u>by direct construction</u>, the shadow theory amplitude is <u>continuous</u>. (It seems to be one of the possibilities considered but then abandoned by Iagolnitzer and Stapp.[4])

The basic reason for things to be this way is that the shadow states can be thought of as an energy-dependent "size" for particles and the corresponding non-locality falls off only by a power law. Such a geometric fall off is at variance with our notion of forces falling off exponentially. It also makes it necessary for us to take account of detectors and other apparati and their modification of the quantum mechanical process with such a long range interaction. The clear but nontrivial declaration of shadow state theories is that under conditions where the interactions are between well-defined particles the transition amplitude to physical states alone conserves probability.

There is one technical point which bears directly on the theme of this conference: the question of reversal of cause and effect. In an indefinite metric theory it is possible to have poles on the physical sheet at points $m_o + i\Gamma/2$ which may correspond to physical channels. With poles on the unphysical sheet at $m_0 - i\Gamma/2$ we have a simple physical picture, namely that of an unstable particle. If we consider the collision of two particles head-on to produce the unstable particle which subsequently reemits these particles their space-time trajectory can be drawn as in Fig. 1.

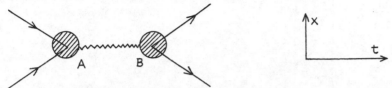

Fig. 1. Collision of two particles propagated by an unstable particle corresponding to a pole at $m - \frac{i}{2}\Gamma$. (After Stapp.[4])

In contrast for the collision in case the pole is on the physical sheet it appears that we have a causality violating sequence shown in Fig. 2.

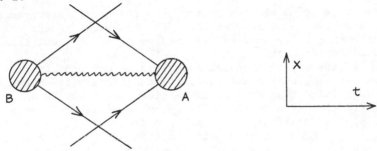

Fig. 2. Collision of two particles advanced by an unstable particle corresponding to a pole at $m + \frac{i}{2}\Gamma$. (After Stapp.[4])

Actually this picture is incorrect; instead what we obtain is the situation depicted in Fig. 3. The dotted portions do not take

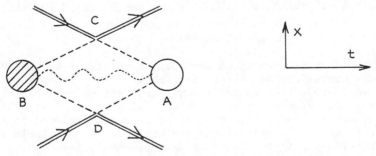

Fig. 3. Collision of two particles in a channel with a pole at $m + \frac{i}{2}\Gamma$.

place at all and the causal anomaly vanishes. We may, of course, think of this as due to a size for the particles. (A size for either one of them will do.) The situation may be illustrated by the "orbit diagrams" Fig. 4a,b. In the second case the particles

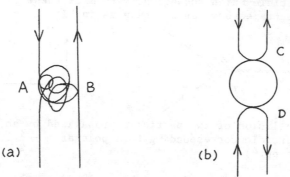

(a)   (b)

Fig. 4. "Orbits" for the particles in cases described in
        Figs. 1 and 3.

never really "meet" each other. The case therefore corresponds to a long range repulsive interaction. The behaviour of the phase shifts through the "resonances" in the two cases bear out this physical interpretation.

Even if there are no causal anomalies we may want to restrict the "size" of the interacting particles to microscopic dimensions. This would require that in Fig. 3 the dimension CD is "small". This entails that AB also is small; in other words the imaginary part $\Gamma/2$ of the "mass" of the unstable pole be "large", that the width be of the order of strong interaction widths. Such large complex masses do not arise from real masses in an indefinite metric theory of weakly coupled satellite quanta like for lepton satellites in quantum electrodynamics or in weak interactions. But in an indefinite metric theory we may, if we choose, start with pairs of complex conjugate masses. We must not lose sight of the possibility, however, that "large sized objects" are compatible with shadow state theory; and we may consider it worthwhile to be alert to the possibility of macroscopic impact parameter collisions. They would appear as disconnected scattering events in a bubble chamber.

WHO NEEDS SHADOW STATES?

Even if we are convinced about the consistency of the physics of shadow states one may raise the question: why bother? Why should we include such an additional complication into the theory? There are two enticing reasons that appeal to me: First, there are a number of assorted particles that have been postulated in particle physics from time to time which have not been found. Included in this "missing particles list" are magnetic monopoles,[13] intermediate vector mesons,[14] tachyons[15] and quarks.[16] Depending on one's prejudices, background, philosophy and friends one or other of these species of particles appear more essential than the others. But they all share the distinction of not only being not discovered but it appears as if they are as plentiful as the unicorn. It is tempting to suggest that some or all these particles are shadow particles. They enter dynamics but are "not really there"! Since my aim has been to display the logical structure of shadow state theories I will not attempt discussing any models in this presentation.

Second, local relativistic quantum field theory inevitably leads to infinities in any direct physical computation from a Lagrangian. As long as one persists in local interaction structures the infinities persist. The only way out seems to be to introduce fields whose fundamental equal-time commutator (or anticommutator) has the opposite sign to the usual theory so that the corresponding propagators differ by a sign.[3] Such "negative weight" fields correspond to fields defined over an indefinite metric space. By suitably blending positive and negative metric fields we can obtain finite relativistic local field theories. But there was always a difficulty with probability interpretation of the indefinite metric state space. The restriction of the real state space R to states containing only positive metric quanta resolves this problem once and for all. I have discussed these questions in detail elsewhere[2,5] and I shall have to be content with that.

RECAPITULATION

In conclusion allow me to state the following: (i) Quantum theory allows a generalization to shadow state theories with a consistent probability interpretation. (ii) Shadow states enter dynamics but do not enter unitarity. (iii) Modified Green's functions of shadow state theory do not allow consideration of autonomous particles in the shadow subspace. (iv) Considerations of shadow states as if the individual particles are propagating autonomously are incorrect; and they lead to incorrect conclusions about the consistency of shadow state theories. (v) Some interactions not carried by physical objects fall off no faster than a power of the distance. (vi) Shadow states are useful to give a consistent interpretation of indefinite metric theories. (vii) If we have the conventional scattering amplitude computed using the retarded Green's function, we can simply obtain the physical scattering amplitude. (viii) Magnetic monopoles, tachyons, or quarks could all be accommodated as shadow particles and this may make them totally unobservable. (ix) Piecewise analyticity is inevitable in shadow state theories.

The question of piecewise analyticity which upsets many is not so far removed from our experience after all. Whenever we look into a mirror we see a causal world, obeying a number of simple laws. For every object placed in front of the mirror we see another one, laterally inverted, in the mirror. If we have a field of illumination the illumination is a smooth function of position. But if we actually proceed to look behind the mirror we see a very different

128

Before the looking glass   Behind the looking glass

(a)   (b)

Fig. 5. A Piecewise "analytic" world view.

physical situation. Until we get to know about mirrors we will have
an unusual world view. What I have presented here is just another
case of a possible piecewise-analytic world view.

REFERENCES

1. H. Hertz, "Principles of Mechanics", Miscellaneous Papers
    Vol. III (MacMillan, New York, 1896).
2. E.C.G. Sudarshan, "Action-at-a-distance", Fields and Quanta
    2, 175 (1972).
3. E.C.G. Sudarshan, "Quantum Mechanical Systems with Indefinite
    Metric I", Phys. Rev. 123, 2183 (1961); "Indefinite Metric and
    Nonlocal Field Theories", Fundamental Problems in Elementary
    Particle Physics, Proceedings of the XIV Solvay Conference
    (Interscience, New York, 1968).
4. H.P. Stapp, "Macrocausality and Its Role in Physical Theories",
    Proceedings of the Conference on Causality and Its Role in
    Physical Theories, Detroit, Michigan (1973); D. Iagolinitzer
    and H.P. Stapp, "Macroscopic Causality and Physical Region
    Analyticity in S-Matrix Theory", Communications in Mathematical
    Physics 14, 15 (1969).
5. E.C.G. Sudarshan, "Shadow and Substance", Proceedings of the
    Powai High Energy Physics Conference, Tata Institute of Funda-
    mental Research, Bombay (1972); to be reported in the report
    of the study group on Project Isabelle, Brookhaven National
    Laboratory, Upton, New York.
6. Compare the discussion in C. Castaneda, "A Separate Reality",
    (Simon and Schuster, New York, 1971); especially pp. 164-5.
7. C.A. Nelson and E.C.G. Sudarshan, Quantum Field Theories of
    Shadow States--I Models, and II Low Energy Pion-Nuclear
    Scattering, Phys. Rev. D6, 3658, 3678 (1972).
8. N.F. Mott and H.S.W. Massey, "The Theory of Atomic Collisions",
    (Oxford University Press, London, 1965); Chapter II.

9. J. Dhar and E.C.G. Sudarshan, "Quantum Field Theory of Inter-
    acting Tachyons", Phys. Rev. D174, 1808 (1968); C.C. Chiang
    and A.M. Gleeson, "S-Matrix for Finite Quantum Electrodynamics
    in Heisenberg Representation" (to be published in Phys. Rev. D).
10. J.-L. Richard, "Formal Theory of Scattering of Shadow States",
    Phys. Rev. D7, 3617 (1973).
11. M.G. Gundzik and E.C.G. Sudarshan, "Some Consequences of a
    Piecewise Analytic Scattering Amplitude", Phys. Rev. D6, 796
    (1972).
12. In the Indian tradition it is said:
        Time is but one thought taken to be many
        Same Time for many thoughts is Space
        Time and Space coming together is Causality
        From them arise the World around us.
13. P.A.M. Dirac, "Theory of Magnetic Poles", Phys. Rev. 74, 817
    (1948).
14. H. Yukawa, Proceedings of the Physico-Mathematical Society of
    Japan 17 (1935); "Models and Methods in Meson Theory", Reviews
    of Modern Physics 21, 474 (1949); T.D. Lee and C.N. Yang,
    "Implications of the Intermediate Boson Basis of the Weak
    Interactions: Existence of a Quartet of Intermediate Bosons
    and Their Dual Isotopic Spin Transformation Properties",
    Phys. Rev. 119, 1410 (1960); see, however, E.C.G. Sudarshan,
    "Nature of Primary Interactions of Elementary Particles",
    Proceedings of the Royal Society 305A, 319 (1968).
15. O.M.P. Bilaniuk, V.K. Deshpande, and E.C.G. Sudarshan, "Meta
    Relativity", American Journal of Physics 30, 718 (1962); G.
    Feinberg, "Possibility of Faster-Than-Light Particles", Phys.
    Rev. 159, 1089 (1967); E.C.G. Sudarshan, "The Nature of
    Faster-Than-Light Particles and Their Interactions", Arkiv
    für Fysik 39, no. 40 (1969).
16. M. Gell-Mann, "A Schematic Model of Baryons and Mesons",
    Phys. Letters 8, 214 (1964).

DISCUSSION

Editor's Note: The following was edited by both the editor and the
speaker from tapes of the question and answer period which followed
the talk.

NEWTON: The scattering amplitude is defined physically either by
means of wave packets, as a relation between a state at time $-\infty$
and a state at time $+\infty$, or else in a time independent way in terms
of what you send in, the beam you sent in, and what comes out.
You've defined in a formal sort of way something that has the
virtue of being unitary but what has it got to do with physics?
In other words, what has this amplitude got to do with any observation of that kind?

A. Let me restate the two ways of defining the scattering amplitude. In the first case you choose a state of the exact Hamiltonian which strongly converges to a preassigned solution of the free
(comparison) Hamiltonian in the infinite past and call it the "in"
state; the corresponding state which converges to the preassigned
state in the infinite future and call it the "out" state. If we
now consider the scalar product of an "in" and an "out" state that
gives the scattering matrix from which the scattering amplitude
can be obtained. Another definition of the scattering amplitude
is as the matrix element of the interaction between the "in" state
and the states of the free Hamiltonian. In my theory I do very
similar things. It is easier to draw the correspondence with the
first method: we simply make a different definition of the "in"
and "out" states! This, of course, reflects the different boundary
conditions employed. It must be stressed that the state χ is an
(ideal) eigenstate of the exact Hamiltonian; and it would be the
state obtained by summing the perturbation series for the eigenstates of the exact Hamiltonian. This is a physical choice (unless
you forbid me to use the word physical in my theory). That means
that when I calculate the scattering amplitude I choose to calculate it using this identification of the states. And if you do
experiments I would urge you to compare your data with the predictions. They may or may not agree but, then, this is to be found
out. The difference comes in, if you like to view it that way,
from the contribution from an elementary scatterer being chosen
differently by the "establishment" and in this theory. This freedom of choice is possible because there are solutions of a free
wave equation which are proportional to a source.

NEWTON: But it's not just proportional to the source. It's also
that a choice of sigma means that it is a very specific choice of
phases at large distances. Because the choice of σ amounts to
a principle value Green's function in that region, which means that
for those shadow particles, the phases have to be just exactly like
you chose them at infinity; which means that the particles that
you don't want to acknowledge as particles nevertheless have to
be controlled by infinity.

A. Exactly, exactly, but I would not control them but instead I
identify the states in which they are controlled! In fact this
will formally come up in Stapp's discussion: I don't want to
anticipate too much of the discussion. If you consider a multi-
particle scattering amplitude, when you "cut it" out you see that
there is something which appears like a singularity, corresponding
to free particles, but it contains incoming and outgoing waves
and therefore a nonvanishing real factor but no imaginary part
corresponding to the shadow intermediate states.

DRESDEN: Am I correct in believing that the fundamental formula
you wrote down is essentially an expression for the T-matrix,
which has the σ's in it?

A. Right.

DRESDEN: But I did not quite understand your claim. Because is the assertion that if I now start computing matrix elements of processes with that new T-matrix (for whatever reasons) no matter what choice I make of σ I will get finite results?

A. No, what I said was that no matter what σ you choose, you get a consistent answer in the sense that when restricted to the nonshadow subspace you get a unitary amplitude. Now, if you want a _finite_ amplitude you start with a _finite_ theory! If the original theory was not finite there is no way of getting one by this method. The only thing that the theory of shadow states enables you to do is to make a finite unitary theory in terms of interacting particles, by introducing indefinite metric, negative probabilities complex mass and other such "unphysical" features into your theory.

DRESDEN: Yeah, but I thought that effectively you had some kind of a compensation mechanism and σ allows you to make, construct....

A. In a sense the only thing that σ does is to allow you to have things which you don't then have to pay for later on. You can have negative probabilities in the virtual states, and it is always the virtual states which create trouble, not the real states.

DRESDEN: I see, so there are really two quite separate elements.

A. Right, in fact I wanted to distinguish between them, but a shadow state need not necessarily be a negative metric state. Even though that is the context in which I developed the formalism.

FADNER: I would like to suggest that you've got a very elegant and particular example of a more general principle. And that is, this of course was stated by you, that the probabilistic interpretation of Ψ is not necessary. In fact it is not necessary in a Heisenberg sense in that we never observe it. The only thing we ever do observe is an interaction probability, and so although it is handy and useful to normalize Ψ through letting the integral of $(\Psi^\dagger \Psi)$ over all space go to one, in many cases at least, this is not necessary. But it is sufficient to give us a good and reasonable theory for things we actually observe, which are things such as scattering. So I would like to suggest that this is a more general principle which is not really stated very often. The second thing I would like to suggest is it's possible that the theory you are presenting on the basis of shadow states, is also sufficient, but not a necessary theory in order to get the thing we are really after, which is a probabilistic interpretation of interactions.

COOKE: Could you compare what you have done with the Gupta-Bleuler ideas in electrodynamics?

A. The celebrated Gupta-Bleuler work, of course, is much more elegant in a certain sense, perhaps because it was inevitable. In electrodynamics, the good Lord said "Let there be light" and then there was light; and then there was Maxwell's equations and you have to make a quantum theory, and the simplest most elegant way of doing it is the formalism developed by Gupta. On the other hand, one also recognizes that the Gupta-Bleuler method does not lead to a <u>finite</u> theory. It's very elegant, but too conventional to be able to eliminate divergences. It is the most beautiful theory of that kind: but it does not do the job of making the theory finite.

COOKE: I know, but I was asking not so much about the questions of the theory being finite....

A. Perhaps you're asking about the negative metric aspect of it? But in some ways the negative metric aspect in the Gupta-Bleuler thing may be stated very simply. (I suppose that neither Gupta nor Bleuler would be very happy with the way I put it: but I hope that both of them are absent from the audience.) If you had Maxwell's equations, you may write down the equations in the form:

$$\partial_\nu F_{\mu\nu} = J_\mu$$

and

$$\partial_\mu A_\nu - \partial_\nu A_\mu = F_{\mu\nu}$$

From these two equations, it follows that:

$$\partial_\mu \partial_\nu A_\nu - \Box^2 A_\mu = J_\mu$$

In the Gupta-Bleuler form, that is with all the four components taken in this particular fashion, we have: $\Box^2 A_\mu = -J_\mu$ and therefore, we have the equation that

$$\Box^2 \partial_\mu A_\mu = -\partial_\mu J_\mu$$

And if you so choose the sources such that the right-hand side is equal to zero, it follows that $\partial_\mu A_\mu$ obeys a free field equation. Now it is up to you whether it should be zero, finite or any number that you want, and it doesn't affect the rest of the problem; it's an independent field so you can choose to have no quanta of this kind to start with. If you start with such a thing then automatically in that theory, all the states that you have, have either zero norm or positive norm. But this is possible only because of the fact that the source that you had here in this particular case was conserved, and therefore this particular combination obeys the free field equation. You have a supplementary condition and in a certain sense the supplementary condition that you have is not that there are no negative norm particles, but in fact, the negative norm particles and the positive norm particles have to come in a certain phase, if they come in at all. That condition

would be propagated for all times. The Hamiltonian consisted of
something which would commute with this condition. This condition
was something which is preserved. In our case the situation is not
so. σ is not a constant of the total Hamiltonian. Consequently,
any supplementary condition that was imposed on σ was going to
affect the time development of the system. σ did not commute with
V and therefore did not commute with the total Hamiltonian. In
Gupta-Bleuler electrodynamics the supplementary condition is something which was consistent with the equations of motion, therefore,
this choice did not affect the problem whatever.

MEYER: Talking about particles, you appear to use the term particle
in some general sense and also in a more specific sense. Now, am I
correct in supposing that when you are alluding to particles specifically, you mean such well established particles as photons, electrons, protons, neutrons, atoms, and so forth?

A. In this discussion, I specifically did not include atoms because I was speaking more in terms of a field theory of fundamental
interaction. But on the other hand, if at the level on which I was
thinking the atom is a single entity, I could include atoms also.
But I also include such objects, if you like, (in solid state)
called 'phonons' which are good, longitudinal photons which are
nonexistent but very important, and a whole lot of other things
like that--rho mesons which are very useful, but don't exist, and
tachyons which don't exist and are probably not useful.

MEYER: When you use the phrase, visible particles, are all of these
particles that we have just been mentioning, are they visible
particles?

A. Let me be a little less rhetorical and a little more precise.
When I said visible particles, I meant a particle which can be
seen. 'Seen' in terms of neutrons being seen or protons being
seen, in the sense that I can detect it. I can talk about a state
(forget all about quantum mechanics) and finally when I am "observing" I observe the particles by direct or indirect means.

MEYER: And therefore everyone of these particles....

A. Everyone of these particles. But there are other particles
like, for example the rho meson, which you do not detect.

MEYER: You mean photons are invisible particles?

A. No, no, rho mesons. Some photons are seen, some photons are not
seen. The virtual photons are not seen. But they are seen indirectly because they are very important.

MEYER: OK. Now in the general sense that you are using the term particle, do you mean that all of these particles are particles of matter, some of them are particles of matter, or none of them are particles of matter?

A. The 'matter' is to be defined in terms of what you want it to be. Let me bring the discussion to a more physical level. Are phonons that are occurring in a Debye solid, particles of matter or are they fiction? For all particle purposes, they are particles of matter, because they do contribute to specific heat and entropy and all such things; they scatter against each other; and yet you know that if you take a bar of iron there are no phonons "inside". Iron is very solid. And yet it is very useful in some contexts to think of it as a phonon gas. I believe that all statements about particles are fiction, just like all statements about material objects. There are no tables! Not even people!!

ROLNICK: When you have these negative metric states taking part in the dynamics, can that lead to an acausal type of preacceleration or any kind of acausal effects.

STAPP: I will discuss that in my talk, so now that I have the floor: Can I just ask something different? You wanted to relate quarks to shadow states or you suggested that perhaps there might be a relationship between quarks and shadow states. You also wanted to make the relationship between shadow states or that the shadow states were not observable and you also made the point that any combination, or any state that had at least one shadow state in it would be a shadow state. That would seem to say that any combination of quark states would be a shadow state, and hence unobservable.

A. No, no. First of all I was not pushing quarks, I am neither for them nor against them. You can have it if you want. I'm suggesting that here are particles which seem to be very useful for certain dynamical calculations but it seems to be a very great embarrassment for certain other experiments because you don't seem to find free quarks. Yet it looks as if there are free quarks participating. I'm saying that one possibility is to consider the three quarks state, the bound state of quarks, to be physical and only these to be physical and the other things to be shadow states. These are energetically separated and therefore you can do this. Now it is not quite the same thing that I said with regard to other particles because I was not thinking of constructing any bound states. The simplest choice is to identify certain particles and call them the shadow particles and anything which the shadow particle touches, also becomes a shadow. But you are quite right. If I wanted to apply this for the three quark state I would certainly not be doing this, I would be making a little more devious choice. I should also mention, while I have the floor, that in principle, especially on the basis of what Professor Newton and various other people said yesterday, if you want tachyons to exist, one way of controlling them--keeping them from becoming a pest--is to say that they are also shadow particles, that they are only responsible for forces and not seen.

SHPIZ: The expansion that was written down by Richard, usually comes from writing a Schroedinger picture, then passing to the interaction picture. If you do that in this theory then, and you start out with _real_ states, won't you then from time development pass into shadow states?

A. Only in the intermediate states.

SHPIZ: Why not in the final states?

A. Because without them everything is alright; so you don't pass into them.

SHPIZ: You can actually eliminate the possibility--you project out when you take the matrix elements.

A. Yes indeed. Look at the expression for the transition amplitude. It satisfies all the requirements that you want to have, but at the intermediate stages of calculation, you have the shadow states also coming in. So you have a whole system which is developing in time; if you are not looking on the mass shell then you have a whole lot of other things. In the intermediate states, you would find a whole lot of these things. The most economical way of interpreting them, in terms of the only observed restricted set of particles is to say that there are energy dependent forces which look as if it is caused by exchange of particles. But there _are_ no such particles.

BELINFANTE: Is that only so in the development from $-\infty$ time to $+\infty$ time or can you also do it between finite times?

A. It is only valid for the total development from $-\infty$ to $+\infty$ because at finite times you get things off the mass shell which are not necessarily energy conserving and then there are effects coming in from shadow state admixtures.

BELINFANTE: Didn't Haller and Landovitz show that, by working rather differently, mathematically, you could (that is in the photon case) work between finite times. I never read that paper in detail, though. They dealt with the same problem in connection with quantum electrodynamics with the longitudinal photons and the usual method, when applied between finite times, leads to unphysical states. But Haller and Landovitz stated in their paper that one could avoid these unphysical states.

A. I suppose that you would avoid the problem if you include non-local interactions; but I am not familiar with the paper to make any useful comments on it.

HAVAS: It seems to me that your work is exactly the quantum ana-logue of what I discussed yesterday, that what you are calling shadow states, is talking somehow in the language of field theory of action-at-a-distance interactions. I think that you stated that much at the beginning of your talk. What was said just at the end was precisely the analogue, say, of talking of a closed system. You have to go from t = −∞ to t = +∞, to come up with a consistent result.

A. That is gratifying.

CAUSALITY AND METRICAL PROPERTIES OF MATTER IN
A TWO-METRIC FIELD THEORY OF GRAVITY

K. Nordtvedt, Jr.
Montana State University, Bozeman, MT 59715

ABSTRACT

If cosmological structure is the result of two metrical tensor fields, causal structure is surprisingly modified with signals traveling faster than light and effects
preceding causes when viewed by certain inertial observers. Metrical structure is seen to be a property of
matter rather than an attribute of the "space-time" four-
dimensional event continuum.

A type of gravitational-cosmological theory is outlined in
which physical signals may travel faster than the speed of light
and the temporal order of cause and effect is to some observers
reversed.[1] These theories contain two dynamical metric tensor
fields -- $g_{\mu\nu}$ and $h_{\mu\nu}$ -- imbedded in a single four-dimensional
continuum of space-time events. Such a class of theories is developed for several purposes:
  1. Astrophysical implications of signals faster than light
would be substantial.
  2. I wish to stress the viewpoint, by the example of such
theories, that space-time metrical structure is a quality of
matter rather than, ontologically speaking, an independent entity
of physical theory.
  3. Finally, I want to explore the radical limits of possible
causal structure within the constraint of rather conservative en-
largement or adjustment of accepted theory.
  The theory is based on an action integral

$$A = \int L d^4 x \tag{1}$$

where the covariant LaGrangian density L is of the form

$$L = L_1(g_{\mu\nu},M) + L_2(g_{\mu\nu},h_{\mu\nu}) + L_3(h_{\mu\nu},M\acute{}) \tag{2}$$

M represents all particle and field variables needed to describe
ordinary matter. M´ represents a possible other type of matter.
This form of LaGrangian guarantees the local Lorentz invariance of
ordinary matter's physical equations and gives matter's response
to gravity according to the form valid in "metrical theories of
gravity",[2,3]

$$T^{\mu\nu}(M) \sim \frac{\delta L_1}{\delta g_{\mu\nu}} \quad \text{and} \quad T^{\mu\nu}\big|_{g\nu} = 0 \tag{3}$$

($|\atop g$ means covariant derivative with respect to the $g_{\mu\nu}$ tensor).

Ordinary matter "sees" only the $g_{\mu\nu}$ metrical field which guarantees the high-precision empirical results of the Eötvös-type experiments[4] and other consequences of Einstein's equivalence principle.

The field equations for $g_{\mu\nu}$ will differ from general relativity, they now depend on $h_{\mu\nu}$ as well as matter's stress-energy tensor $T^{\mu\nu}(M)$. The second metrical tensor field $h_{\mu\nu}$ will have dynamical equations depending on $g_{\mu\nu}$ and possibly a source of another type of matter -- $\theta^{\mu\nu}(M')$. These field equations result from appropriate variation of the LaGrangian density with respect to $g_{\mu\nu}$ or $h_{\mu\nu}$ etc.

Seeking a cosmological, background solution for the metrical fields, we assume a homogeneous, isotropic matter distribution which leads to these fields having the form;

$$g_{\mu\nu} = \begin{bmatrix} 1 & 0 & 0 & 0 \\ 0 & & & \\ 0 & & -S^2(t)g_{ij}^{(3)} & \\ 0 & & & \end{bmatrix} \tag{4}$$

and

$$h_{\mu\nu} = \begin{bmatrix} A^2(t) & 0 & 0 & 0 \\ 0 & & & \\ 0 & & -B^2(t)h_{ij}^{(3)} & \\ 0 & & & \end{bmatrix} \tag{5}$$

with $g_{ij}^{(3)} = h_{ij}^{(3)}$ representing a closed, flat, or open three-space of constant curvature. t is a cosmic time coordinate. Null signals traveling via the $g_{\mu\nu}$ field will have a speed (when observed in the cosmic rest frame)

$$c^2 = 1/S^2(t) \quad , \tag{6}$$

while signals propagating via the $h_{\mu\nu}$ field will have a maximum speed

$$c'^2 = A^2(t)/B^2(t) \quad . \tag{7}$$

The ratio c'/c is determined by cosmological dynamical considerations since the theory's field equations yield coupled time differential equations for $A(t)$, $B(t)$, and $S(t)$.

Consider the interesting case of $c'>c$. Physical interaction between ordinary matter can then proceed through several channels including propagation at the speed c'. Moving matter produces ordinary electromagnetic fields, etc., or dynamical g fields which can propagate to another region of the space-time continuum at the speed c, or the resulting dynamical g-field can produce h-fields which can propagate at the speed c'.

In the cosmic rest frame there are two forward "light cones" restricting all causal connections to the temporal future and preventing the existence of paradoxical causal chain loops. However when viewed by observers moving with respect to the cosmic rest frame unusual features emerge.

Figure la shows (in the cosmic rest frame) a system α at $z=0$ and time $t=0$ emit radiation which is absorbed by another system β at $z=z_o$ and time z_o/c'. Figure 1b shows the same events in a frame moving to the right at speed v. The absorption event in system β occurs at position $z' = \gamma z_o(1-v/c')$ and $t' = \frac{\gamma z_o}{c'}\left(1-\frac{vc'}{c^2}\right)$

with $\gamma = 1/\sqrt{1-v^2/c^2}$. For speeds $v>c^2/c'$ the absorption at β occurs before the emission at α . The wave travels at speed $(c'-v)/\sqrt{vc'/c^2-1}$ from β to α ; effect precedes cause! It can be further shown that this wave carries negative energy when viewed in this moving frame. These observational surprises in causality do not, however, allow for intervention into systems to produce causal chain loops.

There is no metrical property of "space-time" in this theory. Metrical relations are properties of specific measuring devices. Consider two identical length rulers in the cosmic rest frame, one made of ordinary matter, the other of the possible matter M'. If the rulers are both brought to velocity $\vec{v}$ in the universe they will no longer have identical length, their length ratio being

$$\frac{L}{L'} = \frac{\sqrt{1-v^2/c^2}}{\sqrt{1-v^2/c'^2}} , \tag{8}$$

as ordinary matter's physical equations are invariant under the (Lorentz) coordinate transformation

$$X'^\mu = L(c)^\mu_\nu x^\nu , \quad L(c)^\mu_\nu = \begin{bmatrix} \gamma(c) & -\frac{v}{c}\gamma(c) & 0 & 0 \\ -\frac{v}{c}\gamma(c) & \gamma(c) & 0 & 0 \\ 0 & 0 & 1 & 0 \\ 0 & 0 & 0 & 1 \end{bmatrix} , \tag{9}$$

while the physical equations for the matter M' are invariant under the different transformation

$$X^{\prime\mu} = L(c')^{\mu}{}_{\nu}X^{\nu} \tag{10}$$

where c' replaces c. Similarly synchronized clocks of the two kinds of matter at rest in the universe lose synchronization when both brought to velocity $\bar{v}$ in the universe, their relative rate becoming

$$\frac{d\nu}{d\nu'} = \frac{\sqrt{1-v^2/c^2}}{\sqrt{1-v^2/c'^2}} \tag{11}$$

In general $g_{\mu\nu}$ governs only the metrical properties of ordinary matter's measuring instruments while $h_{\mu\nu}$ governs the metrical properties of the other matter's structures. But since "observers" made of each type of matter may "view" the other matter, these observers must conclude there are Machian-type, non-geometrical, physical interactions between the universe at large and the bodies in the universe, these interactions governing the metrical properties of matter, not space-time. We conclude that this conceptual view of the nature of metrical relations, and the more general type of causal structure present in these types of theories, must be asserted also in general relativity where these effects become "latent" rather than actual due to simplifications and resulting invariances in Einstein's classical theory. The proper interpretation of a theory must come from studying the ensemble of theories to which it belongs, not from just studying the specific theory.

$$\alpha \qquad \qquad S = C' \qquad \qquad \beta$$
$$\bullet \qquad \xrightarrow{\hspace{3cm}} \qquad \bullet$$
$$Z = 0 \qquad \qquad \qquad \qquad Z = Z_0$$
$$t = 0 \qquad \qquad \qquad \qquad t = Z_0/c'$$

Figure (1a). A cause (α) produces an effect (β) via a radiation signal traveling at speed $s = c' > c$ in the universe rest frame.

$$\alpha \qquad S' = (c'-v)\Big/\left(\frac{vc'}{c^2}-1\right) \qquad \beta$$
$$\bullet \qquad \xleftarrow{\hspace{3cm}} \qquad \bullet$$
$$Z' = 0 \qquad \qquad \qquad Z' = Z_0\left(1-\frac{v}{c'}\right)\Big/\sqrt{1-\frac{v^2}{c^2}}$$
$$t' = 0 \qquad \qquad \qquad t' = Z_0\left(1-\frac{vc'}{c^2}\right)\Big/c'\sqrt{1-\frac{v^2}{c^2}}$$

Figure (1b). The same events viewed in an inertial frame moving at speed v to the right now has the effect occurring before the cause and negative energy in the radiation for $vc' > c^2$.

REFERENCES

1. A detailed presentation of this type of cosmological-gravita-
   tion theory will be presented soon in the journals.
2. K. Nordtvedt, Jr., Phys. Rev. 169, 1017 (1968).
3. C.M. Will, Course 56 in Proceedings of the International
   School in Physics "Enrico Fermi", B. Berfotti ed., (Academic
   Press, in press).
4. P. Roll, R. Krotkow, R. Dicke, Ann. Phys. (New York) 26, 442
   (1964).

DISCUSSION

Editor's Note: The following was edited by both the editor and the
speaker from tapes of the question and answer period which followed
the talk.

SAPERSTEIN: So what you are saying is that there is a strong coupling and a weak coupling. Because if they were both strong couplings we couldn't make this separation.

A. The separation is of two forms, first of all the separation is
intrinsically in the Lagrangian here, equation (2), OK? Matter can
only see h via the intermediary of g. So it's necessarily a
gravitational phenomenon.

SAPERSTEIN: But that's just like saying, one electron can only see
another electron via the electromagnetic field. Still and all they
can interact pretty darn strongly.

A. In the electromagnetic case?

SAPERSTEIN: Sure.

A. Yes, but we are talking about gravity. You try out examples of
this and you see how it affects gravitational physics. The h-field
only affects matter by influencing the value of the g-field.

SCHLEGEL: Let me ask the question this way. How would you describe
gravitational waves? Would that be with terms of h-metric or the
g-metric?

A. There's both kinds, in general. There are several kinds of
modes, pure transverse modes of the g-field, just like in general
relativity, that propagate without exciting the h-field. There
are pure transverse h waves which travel at c' without exciting g,
and also there are two partly longitudinal modes that involve both
the disturbances in g and h but fortunately still propagate at c
or c'.

SCHLEGEL: But the h-mode would only be excited by the g-mode?

A. Yes, for instance, you typically write down interaction Lagrangians, and you first look at the simplest forms. You find that things like: $R_{\mu\nu}$ (g) and things like R(g) act like the effective sources for the h-metric and vice versa. So we know for instance that if I have some matter under dynamical change, it creates a localized Riemann tensor that's also dynamical and then that can drive radiation or dynamics of the h-metric.

NEWTON: What's the point of calling it a metric, h? You could do that with any arbitrary tensor field, couldn't you?

A. 'Metric' means measurement and g and h still govern the metrical properties of rods and clocks.

NEWTON: Well which does, g or h?

A. Which matter are you talking about M or M'?

DRESDEN: As I understand it you have a ds^2 and a ds'^2. Now, can you tell me what the difference is between them? Experimentally, I have a rod and now I look and that's ds. Now, that presumably involves all the interactions, both the g's and the h's. Right? So how can there be two different ones.

A. But what you've got to analyze is that we actually measure time and space with physical rulers. It is straightforward to show that rulers made out of ordinary matter have their metrical structure determined by the g.

DRESDEN: And they are independent of h all together?

A. Yes, h is only important in determining what g will be.

SAPERSTEIN: g determines what M is going to do. M is coupled to M' since M' is coupled to h, h is coupled to g, g is coupled to M, you throw out all the intermediate steps and say M is coupled to M', unless you have some physically distinguishable characteristic, then it is just names.

A. No, I'm sorry. The point is this: You give me a g in this locality and I will tell you the entire metrical structure of ordinary matter, made up into rulers and clocks. All I have to know is g here, to tell you the geometry and chronometry of ordinary matter. How g got to be what it is, is a total question of what the past history of h is, etc. But you don't have to know what h is here, to know what the metrical structure of ordinary matter is.

SAPERSTEIN: But that is just like saying, given a g here, there is no such thing as an h. The matter someplace else determines what the g is here and that g here determines what the mass here does. There is no need for an h. Why aren't there 59 different h's? What is there <u>physically</u> that tells me what this h is?

A. You can detect it.

DRESDEN: Are you telling me that there are two kinds of geometry. There is something called the blue geometry that is the g one and a red geometry that is the h one. Is that correct?

A. I don't believe in geometry. There are rods and clocks and their metrical structure and relationships. But you are right in that the matter coupled to the h-field, or bundles of h-field energy itself, possess different geometrical properties.

DRESDEN: Good, but you have the ds and the ds'.

A. I just wrote that down. There are two invariants that I can form from neighboring points in the continuum. That's all. That's the only purpose.

DRESDEN: We have a neighborhood notion. There are two types of neighborhoods—good and bad. Or red and blue, right. But they are determined by g and h. They are different.

A. Yes, there are two invariants between ....

DRESDEN: OK, Now what I don't understand is that if you have the neighborhood notion àlà g, right. Why does that not depend on the presence of h?

A. It does.

DRESDEN: Then I don't understand how there can be two of them.

BELINFANTE: Maybe we can say it differently. The h-field is like a meson field which has a very unusual interaction with the gravitational field, like ordinary matter, only differently. So we have ordinary matter coupled with the g-field and the crazy meson field h couples with the g-field in a different way.

DRESDEN: Oh, I see.

A. Suppose I had done it this way. Suppose that I had a g- and h-field and coupled all hadrons to the g-field and all leptons to the h-field. Well first of all, it would have destroyed the precision of the Eötvös experiment. What it would mean for instance is that the geometry of rulers made out of electrons, would turn out to be slightly different than the geometry of rulers made out of hadrons.

What I am trying to stress is that the geometry lies in the material and the fact that we have one geometry which seems to hold for all of matter, is an expression of the fact that all our matter seems to be just coupling to one background tensor field.

ROLNICK: I might just say that it seems to me that g is the metric field and that the h is the way of introducing a different kind of action, than what Einstein had introduced for the action, corresponding to the gravitational field. So that it is a different gravitational theory, but the h is just a particular way that he has chosen.

DRESDEN: Does h have any invariance properties?

A. Yes.

DRESDEN: Why?

A. During cosmological evolution there was substantial mixing of the g- and h-field. We traditionally pick coordinates to make the local background g into the Minkowski form. Then the background h is of the form

$$h_{\mu\nu} = \begin{bmatrix} c'^2 & 0 & 0 & 0 \\ 0 & -1 & 0 & 0 \\ 0 & 0 & -1 & 0 \\ 0 & 0 & 0 & -1 \end{bmatrix}$$

It has an invariance group, too, Lorentz transformations with a different c'.

DRESDEN: You would then agree that the physical reason for that Lorentz invariance and the other one is totally different.

A. No.

DRESDEN: There are no rods in the usual arguments you do....

A. There are rods.

DRESDEN: h-rods?

A. Yes. There are h-rods! Suppose I made a gravitational ruler out of some bodies orbiting the Earth. For instance the radius of the Earth's orbit is one of the rulers that we use.

DRESDEN: Fred almost explained it to me, but now I no longer understand it.

A. If you viewed the Earth's orbit, moving by the Earth very fast you would find the Earth's orbit collapsed due to the Lorentz contraction of (M) rulers. Now suppose that there were some rulers made out of gravitational forces of the h-variety. Other matter held together gravitational by the h-metric, or h-waves themselves. Then, these rulers will contract according to a different Lorentz contraction and in general, if you take two of these rulers in one frame and two in another frame, they will not measure the same intervals in all different Lorentz frames because they are invariant under different....

DRESDEN: But you see, what Fred explained to me is that I must understand by M, that's ordinary matter that's held together by gravity g. And now you have this crazy meson field. That meson field is a law unto itself. It can do whatever it wishes to do. And whatever our experiences are about the gravitational field, they have no particular relevance for this field, h.

A. Except it is coupled to g.

DRESDEN: But now its invariance properties might very well be totally different, in fact in general they would be, from those at M. To somehow just carry over, of course you are allowed to do that, but I see no reason to.

A. If the background h-field is arbitrary, you're right. The invariance group of that matrix is who knows what. But if it is equation (5), you know very well that it is another Lorentz transformation with c replaced by c'.

DRESDEN: You not only want two Einsteins; you want two Lorentz's as well.

WEISS: How do you measure c'?

A. By an experiment with matter here and there, separated by a length L. A collapsing star or something not only emits photons which travel at c, not only emits g-gravitons which travel at c, but emits h-gravitons which travel at c', and therefore the causal influence arrives here at a much earlier time.

WEISS: So you postulate the existence of such h-gravitons.
A. Yes.
DENMAN: Why do you say that the observer will not change the course of events. That is, he will not decide not to send out the signal.

A. Because there is one frame in which you can analyze any experiment which is the cosmic rest frame. All causal signals go forward in time at least faster than the c' light cone. Therefore you can show that any time-like world lines can just influence the future. There is at least one frame in which all causal influence goes forward in time. And by the covariance of the theory you can use this frame to analyze any experiment you choose. It may not be the simplest one.

NEWTON: Covariant with respect to g or respect to h? The first term is covariant with respect to g the last term (that you have erased there) presumably would be covariant with respect to h wouldn't it?

A. There is one set of coordinates. It is covariant with respect to using any set of labels for the continuum to describe the physics. There is one continuum, and covariance, as I understand it, is a statement of the arbitrariness of how you want to label the continuum.

NEWTON: But you happen to have a metric. Therefore, use the metric.

ANTIPPA: I would just like to make a comment. In doing the electrodynamics of tachyons in a two metric theory, there is a condition so that two light cones are tangential instead of coaxial. One physical effect that comes out of the geometry that tachyons obey, a tachyon charge emits hyperbolic electromagnetic waves. This is an example of another two metric theory with a well defined physical difference.

FADNER: Does your theory have the possibilities of tachyons and extraordinary tachyons one going according to the g- and the other according to the h-metric. Does that therefore have the problem of causality.

A. Yes. I haven't worried about tachyons, yet.

FADNER: OK, but others have. You are saying that its possible to get something faster than the speed of light without worrying about tachyons. You still said that you are not going to worry about whether tachyons exist in your separate state of your two kinds of materials.

A. You see no matter what you do here, there is still an ultimate Lorentz group with the highest c'. And therefore, you just call that your ultimate speed, at least in the cosmic rest frame.

FADNER: Are you postulating that or are you saying that it has to come out of your....

A. I'm saying that I don't want to postulate anything a priori. It should come out of some dynamic equations. The theory ought to contain in the cosmology, the dynamics that determines what c' is relative to c. One thing I should comment. The fast gravitons have a possibility of Cerenkov radiation into the slow gravitons, and I haven't yet been able to calculate what the rate is. It would be pretty slow. It is a weak coupling, but energetically you can show that one of these things could Cerenkov radiate.

TACHYONS, CAUSALITY, AND ROTATIONAL INVARIANCE

Allen E. Everett
Department of Physics, Tufts University
Medford, Massachusetts

and

Adel F. Antippa*
Département du Physique, Université du Québec
Trois Rivières, Québec, Canada

ABSTRACT

We extend a previously developed one dimensional causal theory of tachyons to three dimensions. The result is a three dimensional theory of interacting tachyons in which coordinates in reference frames with subluminal relative velocity are related by the Lorentz transformations, and in which no paradoxes involving causal loops can arise. The resulting theory involves a preferred spatial direction and preferred velocity perpendicular to that direction, so that physical laws governing tachyons are not invariant under rotations or proper Lorentz transformations. This lack of invariance should manifest itself even in processes involving only bradyons, to the extent that coupling to virtual tachyons is important. We discuss the limits which experimental evidence on the validity of rotational invariance places on tachyon couplings in the theory and possible additional experiments for searching for lack of such invariance.

I will describe a theory of interacting tachyons in 3 spatial dimensions which is free of causal loops or cycles. That is, the kind of self-contradictory situation in which an event happens if and only if it doesn't happen or doesn't occur in the theory. There are, in the theory, causal sequences, in the sense described yesterday by Professor Newton, in which an effect precedes the cause for some observers. However, these simply mean that a physical law which we have hitherto believed, namely that the cause, in Professor Newton's sense, always precedes the effect doesn't hold in a new class of phenomena, a common situation in physics. They do not imply the presence of logical contradictions. Thus one can at least formulate the theory and confront it with experiment. Presumably one can't even do this if causal loops are not eliminated. Reference frames in the theory with relative speed $|\beta| < 1$ (we use units with $c = 1$) are connected by Lorentz transformations. However, tachyon phenomena single out a preferred direction in space and

*Supported in part by the National Research Council of Canada.

also violate invariance under proper Lorentz transformations. This
implies a limit, and quite a small limit, on tachyon couplings in
order that virtual tachyon effects not lead to violation of rotational invariance (RI) of a magnitude excluded experimentally in
bradyon processes. (We use the term bradyon to refer to particles
with $|\beta| < 1$.)

   We have shown using the extended Lorentz transformations to
reference frames with $|\beta| > 1$ in one spatial dimension developed
by Antippa and also by Parker[1] that one can avoid the problem of
causal loops in the one dimensional case. One finds that a signal
can't return to its starting point in both space and time. I'll
just sketch the argument. The one dimensional transformation
equations are the following:

$$x' = \mu(x - \beta t)(|1-\beta^2|)^{-1/2} \tag{1a}$$

$$t' = \mu(t - \beta x)(|1-\beta^2|)^{-1/2} \tag{1b}$$

where

$$\mu = 1, \quad |\beta| < 1; \quad \mu = \beta/|\beta|, \quad |\beta| > 1 \tag{2}$$

The definition of μ as given in Eq. (2) is required in order that
the transformations obey the group property. x' and t' as given by
Eqs. (1a) and (1b) are real as they must be since they represent
observable quantities in the theory, namely the coordinates of an
event as seen in the primed coordinate frame. These equations imply
that there are only two classes of reference frame (or equivalently,
two classes of particles, since the possible particle rest frames
define the possible reference frames) in the theory. We might call
these class I and class II reference frames, and refer to the two
classes as "ours" and "theirs" respectively, meaning by this that
class I frames have speed less than 1 relative to us while class II
reference frames have speed less than 1 relative to a hypothetical
superluminal observer. All members of a given class have relative
speed less than 1, and have relative velocity of magnitude > 1 with
respect to any member of the other class. Hence particles may be
divided into tachyons, with $|\beta| < 1$ in our reference frames, or
$|\beta| > 1$ in theirs, and bradyons, for which the situation is the
other way around. The transformation equations are completely
symmetric between the two classes of reference frames. If we use
the inverse form of equations (1) for the case that the primed reference frame S' is of class II one finds that if $\Delta t' > 0$ for two
events on the world line of a tachyon then Δx is always > 0 in any
class I reference frame (or Δx is always < 0, depending on the definition of the positive spatial direction); this follows immediately from Eqs. (1) and the fact that, for the case under discussion,
$|\beta| > 1$ and $|\Delta x'| < |\Delta t'|$. Thus if tachyons can only be used to
transmit signals forward in time in their reference frames (just as
bradyons can only be used to send signals forward in time in our
reference frames), tachyon signals can only be sent forward in space
in class I reference frames. It then follows that signals cannot be

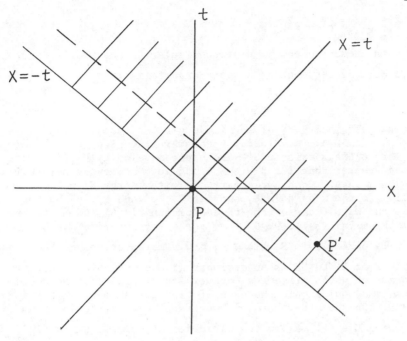

Figure 1. Space-time diagram showing the x and t axes, the
forward and backward light cone, and the regions
which a signal can reach from the origin (point
P) and from P'.

sent around a closed loop in space and time. This can be seen from
Fig. 1. Signals may be sent from the space time point P to any
point in the shaded area. Points in the forward light cone may, of
course, be reached with a bradyon signal, while points, such as P'
for example, may be reached by means of a tachyon signal. However,
from P' signals may only be sent to points above and to the right
of the dashed line, and so no signal may be sent back to P.

To construct a 3-dimensional theory the obvious way to proceed
is to adjoin to Eqs. (1a) and (1b) the usual Lorentz transformation
equations for the transverse components, namely

$$y' = y \qquad\qquad\qquad\qquad\qquad\qquad\qquad (1c)$$

$$z' = z \qquad\qquad\qquad\qquad\qquad\qquad\qquad (1d)$$

However, the resulting set of equations can work only along a given
direction in space. To see this, let us note for example that Eqs.
(1) applied to the energy-momentum 4-vector for the case $\beta > 1$
imply that

$$E'^2 - p'^2 = p^2 - E^2 - 2p_\perp^2 \qquad (3)$$

where $p_\perp$ is the component of $\vec{p}$ perpendicular to the x axis. Let us now consider the sequence of Lorentz transformations

$$S \rightarrow S' \rightarrow S''$$

where S has a velocity β_1 of magnitude greater than 1 along the x axis in S, and S" has a velocity β_2 of magnitude less than 1 perpendicular to the x axis in S', and hence a velocity β_{12} in S of magnitude greater than 1 and with a non-zero component perpendicular to the x axis. Consider a particle with energy E and momentum $\vec{p}$ parallel to the x axis in S, and let E', p' and E", p" be the energies and magnitudes of the three momentum in S' and S" respectively. From Eq. (3) we have $E'^2 - p'^2 = -m_o^2$, where m_o is the rest mass of the particle in S, since $p_\perp = 0$. Moreover, since the transformation from S' to S" is an ordinary (i.e., subluminal) Lorentz transformation which preserves Lorentz scalar products, we would likewise find, using this sequence of transformations, that $E''^2 - p''^2 = -m_o^2$. On the other hand, if we transformed direction from S to S", using Eqs. (1) along the direction of β_{12}, we would find, using (3), that $E''^2 - p''^2 \neq -m_o^2$, since now $p_\perp \neq 0$.

Thus we must agree on a prescription for using the transformation equations (1). To do this we define a set of preferred reference frames having all possible relative velocities (with $|\beta|$ both > 1 and < 1) along a particular direction which we will take to be the x axis and refer to as the "tachyon corridor". Then the allowed coordinate systems in our theory are those that can be reached from the preferred frames by any ordinary Lorentz transformation. One then specifies the procedure for transforming from a general reference frame S to a general superluminal frame S' to be that one carries out the following sequence of Lorentz transformations

$$S \rightarrow S_{pref} \rightarrow S'_{pref} \rightarrow S'$$

One begins by transforming from S to one of the preferred reference frames, S_{pref}, by an ordinary Lorentz transformation with speed $\beta_1 < 1$. One then makes a transformation to a superluminal preferred frame, S'_{pref}, by an extended Lorentz transformation with velocity β of magnitude > 1 and directed along the x axis, using Eqs. (1). Finally one carries out an ordinary Lorentz transformation with speed $\beta_2 < 1$ to S'.

This procedure is well defined. That is to say, it is independent of the choice of the preferred frames, since the 1-dimensional transformations, Eqs. (1a) and (1b), form a group. Causal loops are avoided in the same way as in the 1-dimensional case. Once again, a signal can't be sent around a closed loop back to the same point in space-time (or, indeed, to any point with the same values of x and t as the initial point) as one again has that $\Delta t' > 0$ implies $\Delta x > 0$ if S and S' have superluminal relative velocity. This, incidentally, again shows that the x axis is a preferred direction in space. In fact, the price we've paid to avoid the possibility of causal loops in the theory is the introduction of a preferred spatial direction, that of the tachyon corridor. There is also a preferred velocity, namely that of the preferred reference frames, in the plane perpendicular to the tachyon corridor.

The existence of a preferred direction will show up even in processes involving only bradyons because of virtual tachyon effects. As a result there are strong limits on the possible magnitudes of tachyon couplings to bradyons in the theory which result from the consistency of existing experimental data with the validity of rotational invariance. The best experimental tests of RI appear to be within the domain of nuclear physics. One knows of nuclear γ-decays with rates suppressed by factors of the order of 10^{-18} by angular momentum selection rules. This implies that the amplitude for the admixture of nuclear states of different angular momentum due to the emission and reabsorbtion of virtual tachyons, as in the diagram in Fig. 2, must be $\leq 10^{-9}$.

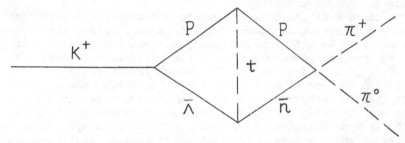

Figure 2. Possible Feynman diagram involving the exchange of a tachyon, t, which could contribute to K decay.

Therefore $g^2_{NNt} \leq 10^{-9} g^2$ or $\leq 10^{-6} \alpha$, where g^2_{NNt} is the tachyon-nucleon coupling constant, g^2 the usual strong interaction pion-nucleon coupling constant, and α the fine structure constant. The cross section for the production of such tachyons would be at most about 10^{-4} μb, too small to have been observed in any of the experimental searches for tachyons,[2] especially since the small limits on the coupling strengths imply, of course, that tachyons in the theory will not only be very difficult to produce but also very difficult to observe directly. The validity of angular momentum selection rules in β-decay puts similar limits on tachyon-nucleon couplings. It also implies that the product $g_{pnt} g_{e\nu t}$, where g_{pnt} and $g_{e\nu t}$ are the proton-neutron-tachyon and electron-neutrino-tachyon coupling constants, must be $\leq 10^{-9}$ G, where $G \approx 10^{-11} g^2$ (in reasonable units) is the usual weak interaction Fermi constant. Otherwise β-decay would occur through virtual tachyon processes at a rate which would lead to observable violations of the β-decay angular momentum selection rules.

One place where the violation of RI predicted by the present theory might show up is in K-meson decay if tachyon couplings are such that virtual tachyon exchange can lead to the occurrence of processes in which there is a strangeness change of one unit. For example, suppose that both g_{ppt} and $g_{\bar{n}\bar{\Lambda}t}$, the tachyon coupling constants to 2 protons, and to an antineutron and antilambda, respectively, are both non-zero, and that their magnitudes satisfy g_{ppt} $g_{\bar{n}\bar{\Lambda}t} \sim G \approx 10^{-11} g^2$, then, for example, the process described by the Feynman diagram in Fig. 2 could compete with the ordinary weak interaction in the decay of the K^+. This would result in a spatial anisotropy in the decay of the K^+ in its rest frame even though it has spin zero. One would have to be careful in looking for this effect in, e.g., a bubble chamber experiment, as in pictures taken at different times a given direction in the bubble chamber corresponds to different directions in space due to the earth's rotation. Since the effect would be washed out unless this is allowed for, it seems likely that the anisotropy, even if it should exist, would not thus far have been detected.

One can ask if there is any way to understand the existence of a preferred direction in space if the tachyon corridor should ever be discovered. We offer one extremely speculative possible explanation. We note that bradyons move in the positive t direction along their world line, which, in the "big bang" theory could be defined as away from the position in time of the big bang. Since, in the present theory, tachyons always move in the positive direction in space along the tachyon corridor, this suggests the possibility that the tachyon corridor might be the direction in space away from the point of the big bang. This means, of course, that we would be dealing with coordinate frames which were curved over cosmological distances. We have made no effort to work out this idea in detail, and offer it only as a speculation.

REFERENCES

1. A.F. Antippa, Universite du Quebec a Trois Rivieres, Report No. UQTR-TH-1 (1970) (unpublished); L. Parker, Phys. Rev. 188, 2287 (1969); A.F. Antippa and A.E. Everett, Phys. Rev. D4, 2198 (1971); A.F. Antippa, Nuovo Cimento 10A, 389 (1972).
2. T. Alvager and M. Kreisler, Phys. Rev. 171, 1357 (1968); M. Davis, M. Kreisler, and T. Alvager, ibid. 183, 1132 (1969); C. Baltay et al., ibid. D1, 759 (1970); J. Danburg et al., ibid. D4, 53 (1971); P.V. Ramana Murthy, Lett. Nuovo Cimento 1, 908 (1971).

DISCUSSION

Editor's Note: The following was edited by both the editor and the
speaker from tapes of the question and answer period which followed
the talk.

DRESDEN: I think that if you take your basic assumption that there
is a fundamental direction therefore the angular momentum in some
sense breaks down, then you don't really need to go through this particular process but a process I would suggest you look at--someone
look at--is the decay of the long-lived K-meson into two π-mesons,
because that is the one which allegedly breaks T, time reversal
invariance. Now in the arguments which lead to that conclusion,
there is a very important argument that the states which one has,
which are the states of π⁺ and π⁻, that there you have rotational
invariance. Now if you tell me that that is not so, you see, then
you could get an admixture of a state, L = 2 and L = 1, and then you
might get it. So in a way, what you ought to look at is the angular
distribution of the π⁺s and π⁻s. That is a straight case. But
then what you would have done, you would then have somehow at least
connected it without any further coupling condition.

A. You still need a diagram....

DRESDEN: I have the K long which should go only into three π-mesons;
it does go into two mesons. Now you can say, why is that so. Well,
if you just write totally conventional strong couplings, you find
that this process cannot take place unless you break T-invariance.
Now the argument for that is that you know something about the
angular momentum state of the outgoing π-mesons. That assumes
that you have strict conservation of angular momentum. You would
say if somehow that decay takes place relative to your preferred
axis in a funny way, then that need not be so. That therefore,
might provide you, possibly, with an explanation of this particular
fact. It is not dependent on any additional couplings of the
tachyons themselves.

A. Well, except that tachyons can contribute to the decay again
only by a diagram of the same form.

DRESDEN: They don't contribute to the decay. It is a straight
statement of the fact that there is a preferred direction in space.

A. There is a preferred direction in space only to the extent that
tachyon couplings are important. If the tachyon coupling constant
is 10^{-50}. OK, there is a preferred direction but you will never
see it, I think. I think you still have to have a diagram of the
same kind, which has to be of at least comparable magnitude to the
usual one, to see any breakdown of the symmetry. I think it is an
interesting question whether the theory might not lead to an explanation of apparent T noninvariance in K° decay. I haven't really

looked at it.

FADNER: You are led to the conclusion that--well not a conclusion--but a postulation of the preferred direction in space in order to get rid of causality problems when going through tachyons and back--through tachyons to the reciprocal frame and back or the symmetrical frame and back and you said that if the tachyon interaction is down by a factor of 10^{-50} or something, then the preferred frame wouldn't be detected. I don't think that that follows at all. You could argue that. I need to talk about a preferred frame if you can go from a regular frame to a symmetrical frame. The preferred direction could still exist. You can use this as an argument for it. I'm trying to get back to what Max said here. It may exist, anyway. Even though you don't have a very strong coupling.

A. It may exist. All that I am saying is that as a practical matter, if you look for an asymetry in the decay of the long lived K into two π's, you will be able to find it only if the tachyon coupling is strong enough that processes involving virtual tachyons can compete with the ordinary weak interactions. You can go to the absurd limit where tachyons exist, but they don't couple at all with bradyons and still, in some sense, there is a preferred direction, but obviously you won't detect it experimentally.

DRESDEN: I hate to belabor this point, but that process--let's assume, for the moment a strict T invariance and all the couplings are usual couplings, then that process won't go at all, period. If you see it then it is therefore presumably a manifestation of the--you therefore could save T-invariance if that is what you wish and it would cost you....

A. It would cost you rotational invariance.

ROLNICK: That process is really CP violating--you have to invoke the CPT theorem to make that a T-violation. No direct T violations have ever been found.

A. There is some fairly decent evidence in $\gamma+p \rightarrow \pi+n$ or backwards. I would like very much to think that is not there.

DRESDEN: I don't like those experiments.

SCHLEGEL: I didn't understand your point about the Big Bang Theory and preferred direction. What was the support theorem?

A. It's support is very weak. It is no more than suggested. Bradyons appear to move always in the +t direction along their world line. Now here's a theory with a spatial direction and tachyons appear to move always in the +x direction along that spatial direction which is in fact a reflection of the fact that tachyons in their reference frames move in the +t direction. This

then just leads me to speculate on the possibility that that preferred direction in space might be the direction away from the
origin of the Big Bang, besides that is the only thing that I can
think of in the physical situation that might possibly give an explanation for the existence of a preferred direction. It is a
little bothersome that nature should pick that direction in space
with no apparent cause.

SCHLEGEL: It would not be spherically symmetric.

A. But there was presumably a point in space and time at which
the....

NEWTON: You still have to have a preferred direction away from
that point in order to talk about your corridor.

A. I see what you are saying.

ROLNICK: Maybe, it's a spontaneously broken symmetry.

INDETERMINISM, TIME ARROW, AND PREDICTION*

F. J. Belinfante
Department of Physics, Purdue University
Lafayette, Indiana 47907

ABSTRACT

Life in general, and man's memory in particular, determine a preferred direction in time ("time arrow"). It is made plausible that the time direction in which the entropy increases is this same time arrow.

In quantum theory, wave functions describe not real (single) physical systems, but ensembles of systems. Reduction of wave packets corresponds to treating a system as a member of a new ensemble. Similarly, splitting an electromagnetic field into IN-field plus retarded field, or into OUT-field plus advanced field, corresponds to different ensembles describing a scattering on which only partial information is available. Our desire to ascribe a source to any light observed is the reason for usually preferring retarded fields over advanced fields. Independent of the fact that retardation of light waves is shown experimentally by Fizeau's measurement of the velocity of light, Afterthought One shows that an eye-object interaction by advanced fields would violate the second law of thermodynamics. This would relate the time arrow of electromagnetism to the one of life. A puzzle remains when this is applied to light traveling between stars and interstellar dust.

The time arrow of cosmology (time direction of expansion of the universe) is sometimes related to the time arrow of life by an extension of Olbers' paradox. An example worked out in Afterthought Two shows that this generalization of Olbers' paradox is not justified. Afterthought Three contains some speculations about possible correlations between the time arrows of cosmology and of life.

The time arrow of quantum theory is determined by the fact that quantum theory in general predicts future probabilities, and that it can retrodict (postdict) probabilities of the past only under specific conditions that often are not satisfied. This asymmetry becomes more understandable when quantum theory is applied to sequences of nonideal (irreproducible) measurements separated by the use of gadgetry that prepares chosen initial conditions for the next measurement. The entering here of choice indicates a relation between the mind with its built-in time arrow, and the choice of the ensembles to which quantum theory then is applied objectively.

*Supported by NSF Grant # GP-29786

We shall now return to the style of yesterday's meeting, talking about generalities. The only difference is that I will be thinking in terms of quantum theory instead of classical theory. Therefore the word <u>Indeterminism</u> in the title of this talk.

THE ARROW OF TIME

I consider the time arrow to be given primarily by the direction in which the upper time limit of events in our memory changes from instant to instant. Our memory of events in time has some similarity to our view of space as we are rowing, looking backwards as we progress. People sometimes say that <u>time</u> <u>passes</u>. This picture of the rower suggests that it may be better to say that <u>we</u> <u>pass</u> <u>time</u> toward larger <u>t</u> values.

Various parts of physics seem to show time arrows. Some of them, but not all of them, seem to be a direct consequence of the time arrow in our memory or knowledge of events.

THE SECOND LAW OF THERMODYNAMICS

Consider the statistical derivation of the Second Law of thermodynamics, according to which the entropy of an adiabatic system increases if it can. Boltzmann's <u>H</u>-theorem would suggest that it would increase toward the past as well as toward the future. If at the present we have a state of which we know that S is not maximal, and we have to guess what S will be in the future, because of our lack of knowledge of the future we'll obviously guess the most likely state, which is the state of maximum entropy. Thus, S will necessarily increase from the present specific state of the system known to us, to the state we predict by lack of knowledge.

As to the past, we might happen to remember an even more specific initial state from which the present state developed; but, if we <u>don't</u> have knowledge of the state which actually preceded the present one, we usually are ashamed of admitting it, and we tend to <u>assume</u> that the unlikely state found at $t = 0$ developed from an even <u>less</u> likely state at $t < 0$, so that we may maintain the validity of the Second Law of thermodynamics for the past as well as for the future, even in absence of a logical proof.

A <u>retro-person</u> whose memory would increase toward our past (living in the opposite time direction) would, no doubt, reason in a time-reversed way, and would claim that S would increase toward our past.

RETARDED POTENTIALS

I want to say something about retarded potentials in Maxwell's theory, but, because I consider fields here from the quantum-mechanical point of view, I shall first say a few words about quantum theory in general.

QUANTUM THEORY

Maxwell fields may be considered to describe ensembles of many-photon states in the classical limit of a large and ill-determined number of photons. How ensembles of one-photon states or of few-photon states can be described by wave packets is discussed in Bohm's quantum electrodynamics of 1951, which is discussed in Appendix H of Part Two of my book A Survey of Hidden-Variables Theories, which will be published by the Pergamon Press and is now in the page-proofs stage. (Chapter 2 of Part Three of that book relates Bohm's theory to the methods of Akhiezer and Berestetskii.) I will not discuss such details today.

Note, however, that waves in quantum theory never describe a single particle or a single system of several or many particles. Waves and state vectors describe not the actual single events taking place in nature, but they describe ensembles of many possible forms of such events that have something in common. They are used to describe systems for which we mainly know just this common property. Thus, like the canonical ensemble is the proper statistical model for a thermodynamical system of which we know the temperature of the heat reservoir with which it is in thermal equilibrium, without knowing the exact energy of the system, similarly the ensemble described by the eigenfunction a of the operator for an observable A is the proper statistical model for a system prepared to be in a state in which we know A to be equal to the eigenvalue corresponding to the eigenfunction a.

This morning, Stapp mentioned that there exist different interpretations of quantum theory, in which state vectors are assumed to have a more absolute meaning in terms of reality as it occurs in nature, but he mentioned that this kind of interpretation leads to difficulties. I would go a step further, and would say that it signals a misunderstanding of the fundamentals of quantum theory. If state vectors are to predict probabilities, they must be properties of entities for which probabilities have numerical values. That is, they should be properties of ensembles. A state vector cannot be a property of a single system describing its present state, in as far as there is for such a single event by itself no such thing as a relative frequency. (If we say that a is the property of a single system on which the observable A has been measured, we really want to say that a describes the ensemble of all systems for which A would be the same, irrespective of possible differences between the individual systems. Therefore we are not surprised that a following measurement of a complementary observable B will not always give the same result for systems in the state a.)

160

Suppose that for predicting probabilities of future events for systems prepared in a certain way we use the state vector of the ensemble of systems thus prepared. From this ensemble we may pick the subensemble of those systems for which some following measurement gave some particular result. (For instance, those that traverse a slit in some screen.) As we thus have prepared a new ensemble, it is logical that it will be described by a new state vector. This choice of a new ensemble with a new state vector is called a reduction of the state vector. This does not violate the Schrödinger equation at all. (The state vector that keeps evolving from the old one by the Schrödinger equation remains valid for the old ensemble, if we care for it anymore.) Reduction means starting the consideration of a new ensemble with new initial conditions. Nobody can forbid us to consider new ensembles. One and the same physical system may be considered part of the old ensemble and of the new subensemble. The choice of ensemble (and therefore the choice of state vector) we make will depend on circumstances and on the purpose we have in mind. As the purpose changes, we should not see any objection in changing the wave correspondingly. Thus, there sometimes is more than one way in which we may want to describe a particular system, and we would choose the manner which is most convenient.

IN AND OUT FIELDS, RETARDED AND ADVANCED FIELDS

If we know a given beam of photons incident upon a given target, we assume the scattered wave to be spherical, because we cannot predict where the photons individually will be scattered. Instead of using a single spherical wave, we could perhaps use a mixed state with probabilities for plane waves traveling in all directions, but it would be more cumbersome and would not so well describe the location of the target if we would make each wave infinitely wide. If, however, there is a screen with a hole as shown in figure 1, and the hole is wide enough to minimize diffraction, at a distance far enough away behind this hole the subensemble of the photons that do pass through the hole may again approximately be described by a plane wave (sideways limited), which might describe the incident beam on a second

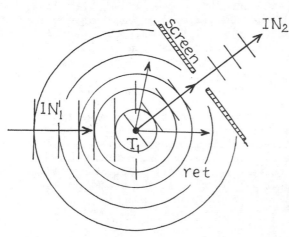

Fig. 1. Plane incident wave and spherical retarded scattered wave.

target. The reduction of the ensemble of <u>all</u> photons from the first target, to the subensemble of photons that traversed the hole, thus naturally is accompanied by a reduction of the wave describing first the entire ensemble, to the wave later describing the subensemble.

Another time, we may detect a photon arriving at the detector D from the scattering target T, without knowing from where a photon had been incident upon T. We might describe this by an outgoing wave in the direction TD, and a spherical wave converging upon T which might be the <u>advanced</u> field of T. This shows that there is nothing that forbids the existence of advanced fields, but that the sum of an advanced field and an OUT field simply describes a different physical situation than an IN field with a retarded field. We choose the description that best fits the problem at hand. (As stressed in Bohm's quantum electrodynamics, the best description of a physical situation does not use plane waves to describe photons, but uses wave packets of finite size.) (See Fig. 2)

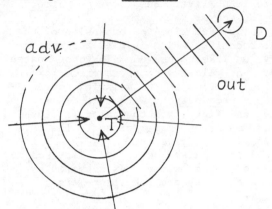

Fig. 2. Advanced field spherically condensing onto scatterer for explaining a wave going out in an observed definite

THE NOCTURNAL SKY

When at night we look at the sky, we notice bright spots, indicating a flow of photons that might be described by nearly plane waves reaching us from many different directions. Plane waves would be solutions of the homogeneous Maxwell equations. Astronomy, however, refuses to believe that the light observed would be a solution of the homogeneous equations. Instead, astronomy claims that the waves reaching us are parts of retarded fields emitted by stars.

Why do we say that we here have retarded fields from the stars exciting our eyes, and why don't we say that these are advanced fields of our excited eyes? The reason is that we do not see the light coming in from <u>all</u> directions. There are dark regions on the sky, and that would not be described by the advanced field of our eyes. Again, the field is chosen to fit the observations.

See, however, <u>afterthought # one</u> at the end of this paper, about correlations between what the observations may be, and the arrow of our memory.

THE EXPANDING UNIVERSE

In the solution to Olbers' paradox, it is reasoned that, if the universe were of constant size (relative to the size of molecular systems), the entire sky should be as bright as the sun, and we would have burnt up. As we are alive, the universe must be expanding. If the universe were contracting, the same reasoning would make us believe that the sky would be even hotter than the sun looks from here, and we would be even less alive. See, however, after-thoughts # two and # three at the end of this paper.

How does relativistic cosmology describe the expansion of the universe, and the cosmological redshift, which is used for solving Olbers' paradox? In cosmology, we like to use comoving coordinates. These are scaled in such a way that galaxies on the average (or if they have no peculiar velocities relative to the galaxies surrounding them) are located at points x = constant. With this description of the space through which the light from the galaxies reaches us, the redshift observed would not be a Doppler effect. The relativistic explanation of the redshift is different.

According to general relativity and the cosmological principle, the invariant line element or proper-time element as measured by meter sticks or atomic clocks is given by the Robertson-Walker metric given by

$$c^2 \, d\tau^2 = c^2 \, dt^2 - R(t)^2 \frac{dx^2}{(1 + \tfrac{1}{4}k \, r^2)^2} \, . \tag{1}$$

The propagation of light waves from the galaxies toward us is described by the vanishing of this expression, so that $c \, dt = -R(t) \cdot dr/(1 + \tfrac{1}{4}k \, r^2)$. If we define the cosmological time coordinate $\underline{t}$ by $\underline{dt} = \underline{dt}/R(t)$, and we define a new cosmological radial coordinate $\underline{r}$ by $\underline{dr} = \underline{dr}/(1 + \tfrac{1}{4}k \, r^2)$, we find that the light from the galaxy at $\underline{r}$ = constant travels toward us according to

$$c \, \underline{dt} = - \, \underline{dr} \tag{2}$$

on this cosmological scale at constant speed $\underline{c}$. So, as shown in the $\underline{t}, \underline{r}$-diagram (Fig. 3), there is on this scale no Doppler effect, and the period $\underline{dt}$ of the light wave as observed now is equal to the period $\underline{dt}$ emitted from the galaxy. Then, however, on the time scale measured by atomic clocks, the ratio of the periods $\underline{dt}$ as observed and as emitted is equal to the ratio of the values of $R(\underline{t})$ at these two times. Since $R(t_{now})/R(t_{past}) > 0$, it follows that on the atomic time scale the period of the light observed is larger than the period that was emitted by the galaxy. This is the cosmological redshift.

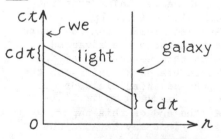

Fig. 3. Space-time diagram on a cosmologically defined scale, showing on that scale a lack of Doppler effect as light from a comoving galaxy reaches us.

The time arrow of cosmology is given by the time direction in which $\underline{R(t)}$ increases. According to the usual reasoning about Olbers' paradox, we would burn up if the time arrow of cosmology were opposite to the time arrow of our life. (See, however, <u>afterthought # two</u>.) I haven't really thought through, yet, what would happen to a <u>retroperson</u> living backward in time in our own expanding universe. (See, also, <u>afterthought # three</u>.)

THE TIME ARROW OF QUANTUM THEORY

The question has been asked why quantum theory should be good for <u>predicting</u> probabilities, and why not for <u>postdicting</u> (or <u>retro</u>dicting) probabilities. Aharonov, Peter Bergmann, and Lebowitz in 1964 have tried to answer this question for consecutive <u>ideal</u> measurements, but it is more interesting also to consider sequences of <u>nonideal</u> measurements interspersed by the application of <u>gadgetry</u>.

By <u>ideal</u> measurement I mean what <u>Von Neumann</u> called a measurement, but what for instance <u>Ballentine</u> would have called a <u>preparation of state</u>. It is the measurement of a conserved quantity that is not perturbed by the measurement, so that a repetition of the measurement would give the same result again. It is the kind of measurement for which the <u>reduced</u> state vector after the measurement is obtained from the <u>initial</u> state vector by a projection and renormalization, to become a normalized eigenfunction of the operator of the observable measured.

Let our ensemble consist of $\underline{N}$ sequences of measurements of observables $\underline{A}$, $\underline{B}$, $\underline{C}$, ... In N_a of these sequences, let the first measurement result in the eigenfunction a of $\underline{A}$ as the reduced state vector. Out of these N_a, let N_{ab} show the eigenfunction $\underline{b}$ as the result of the measurement of $\underline{B}$. Then, the predictive postulate of quantum theory is

$$N_{ab} = N_a \, (a,b)^2, \tag{3}$$

where $(a,b)^2$ is an abbreviation for $|\langle\underline{a}|\underline{b}\rangle|^2$. In the original ensemble, the number of times that b is found is

$$N_b = \sum_a N_{ab} = \sum_a N_a \, (a,b)^2. \tag{4}$$

The <u>predicted</u> probability for $\underline{b}$ when $\underline{a}$ was given (that is, in the subensemble of N_a sequences) by

$$P^a_b \equiv N_{ab}/N_a = (a,b)^2. \tag{5}$$

The question is whether we could <u>also</u> postulate a similar probability for <u>postdiction</u> of $\underline{a}$ in the ensemble of causes in which the later result $\underline{b}$ is given. That is, how about the validity of

$$P^b_a \equiv N_{ab}/N_b = (a,b)^2? \tag{6}$$

Actually, P^b_a is given by (3) with (4). This yields (6) in the <u>special case</u> that in the original ensemble <u>all</u> N_a <u>would be equal</u> (=N/n, if Hilbert space is n-dimensional). In that case, I call this initial state <u>garbled</u>. Then, N_a can be factorized out of the sum in (4), and

$$\sum_a (a,b)^2 = 1 \tag{7}$$

on account of the completeness of the set {a} of possible results of the measurement of A. Equation (4) then gives $N_b = N_a$, so that the ensemble is still garbled at the time of measurement of B immediately following the ideal measurement of A. Then, obviously, (6) follows from (5), in this special case.

Some people, however, have tried to apply (6) more generally, like (5) is generally assumed to be valid. Such misuse of (6) leads to absurd results, so there appears here an asymmetry in time which may be called the time arrow of quantum theory.

THE TIME-SYMMETRIC PROBABILITY POSTULATE

For three consecutive ideal measurements, the number of sequences of results a, b, and c is in predictive quantum theory given by

$$N_{abc} = N_a (a,b)^2 (b,c)^2 \tag{8}$$

For given results of the initial and of the final measurement, the probability for the result b of the middle measurement is

$$P_b^{ac} \equiv N_{abc}/N_{ac} = (a,b)^2 (b,c)^2 / \sum_b (a,b)^2 (b,c)^2. \tag{9}$$

This was derived here as a theorem following from the predictive postulate (5). Aharonov, Bergmann, and Lebowitz asked themselves whether one could not base quantum theory primarily upon the more time-symmetric equation (9) as a postulate replacing (5), or what one would have to postulate additionally for retrieving the conventional theory. Then, the question is why one could not additionally postulate the time-reversed instead, leading to (6) instead of to (5).

They also generalized (9) to a similar postulate for longer sequences of consecutive ideal measurements.

They showed that the predictive postulate (5) could be retrieved if they assumed that there would exist some formula for N_{ab} in terms of N_a only, independent of the probability of later events. Then, N_{ab} would be the same as what it would be if some state in the far future were garbled, and from this, (5) could be derived.

However, they did not answer the second question, why one could not make the time-reversed assumption, from an apriori point of view They just remarked that the time-reversed theory could not generally be valid, as in most cases it disagrees with experience.

GENERALIZATION OF THE AHARONOV-BERGMANN-LEBOWITZ THEORY

The theory just outlined is in need of two generalizations. In the first place, most measurements are nonideal. In the second place, between measurements, one uses gadgets like screens with holes, magnetic fields to bend beams, and whatnot, to prepare the state for the next measurement.

If the state ever was a pure state, it becomes a mixed state by interaction with apparatus in mixed states. I do assume that after each measurement we can single out the object on which the measurement was performed (or which replaces this object, when absorptions or transmutations take place), and that it is possible to consider a reduced state of this object. I just wrote two papers, both still in the referee stage, one about this generalization of the A.B.L. theory, and one about the question of why it should be possible to describe the object after a measurement by a reduced state. (A few preprints are still available.) Talking just about the reduced states, the density matrices of the mixed states are diagonalized, and we can use classical language talking in terms of probabilities, like in the A.B.L. theory.

I shall use primes on the states prepared by the gadgetry which are incident for each measurement, while the unprimed states are the eigenstates corresponding to the possible results of a measurement. The transition probabilities $(a,b)^2$ then are replaced by $(a',b)^2$. The $\{b\}$ are still complete sets, but, since the $\{a'\}$ are not, we have this time

$$\sum_b (a',b)^2 = 1, \qquad \text{but} \qquad \sum_{a'} (a',b)^2 \neq 1. \qquad \text{(10a-b)}$$

If $\underline{A}$ was the quantity last measured before $\underline{B}$, the effects of the gadgetry preparing states $\underline{a}'$ out of the results $\underline{a}$ of the measurements may be described by the transformations

$$N_{aa'} = N_a t_{aa'}. \qquad \text{(11)}$$

Conservation of probability yields

$$\sum_{a'} t_{aa'} = 1. \qquad \text{(12a)}$$

However, in general,

$$\sum_a t_{aa'} \neq 1. \qquad \text{(12b)}$$

The difference between (10a) and (12a) on the one hand, and (10b) and (12b) on the other hand, now describe the asymmetry in time in conventional quantum theory.

Instead of (9), generalized to a longer sequence of measurements, we now have

$$P^{z'e}_{aa'b...dd'} \equiv N_{z'aa'b...dd'e}/N_{z'e}$$
$$= (z',a)^2 t_{aa'} (a',b)^2 ... t_{dd'} (d',e)^2/\sigma_{z'e} \qquad \text{(13)}$$

with

$$\sigma_{z'e} = \sum_a \sum_{a'} \sum_b ... \sum_d \sum_{d'} (z',a)^2 t_{aa'} (a',b)^2 ... t_{dd'} (d',e)^2. \qquad \text{(14)}$$

Again, in predictive quantum theory, (13) is a theorem. Again, we might ask what additional postulates are needed, if we would replace the postulate (5) with $\underline{a}'$ instead of $\underline{a}$, by the more (but not entirely) symmetric (13) as the fundamental probability postulate. In the paper that is still being refereed, I show that predictive quantum theory can be retrieved if we assume (or if someone would prove) that there exists a complete set $\{\underline{f}\}$ of possible results of some possible future measurement $\underline{F}$ such that, for two consecutive sets of prepared states $\{\underline{a}'\}$ and $\{\underline{b}'\}$, we simultaneously would find the same value for all $\sigma_{a'f}$ and for all $\sigma_{b'f}$. This would mean that for arbitrary probability distributions over either the $\underline{a}'$ or $\underline{b}'$ as initial states, the quantity $\underline{F}$ in the final state would be garbled. The whole purpose of this is to make simultaneously the $\sigma_{b'f}$ and the

$$\sigma_{a'f} = \sum_{b\,b'} (a',b)^2 \, t_{bb'} \, \sigma_{b'f}$$

independent of $\underline{f}$. It is then possible to derive the predictive postulate from it.

Contrary to the case of merely ideal measurements, in this generalized case a garbled case due to the gadgetry need not stay garbled.

Because of the lack of time symmetry illustrated by (10a-b) and (12a-b), it is now in general impossible to replace the above additional postulate about a final garbled $\underline{F}$, by a time-reversed assumption about a preceding garbled result of measurement of some quantity $\underline{U}$. It is possible to choose the gadgetry in such a way that this assumption would be in conflict with the probability distribution in effect after the gadget has been applied.

In particular, if the incident states $\underline{z}'$ preceding the first one of two consecutive measurements $\underline{A}$ and $\underline{B}$ are not garbled, it is now impossible to make the postulate that would be needed for validity of postdiction of $\underline{A}$ from $\underline{B}$, even in case $t_{aa'} = \delta_{aa'}$. Therefore, validity of the predictive theory used in eqs. (10)-(12) precludes in this particular case the validity of a postdictive rule.

For more detail, see the paper I wrote on Retrodiction in Quantum Theory, (submitted for publication) which not only gives proofs of the statements which I made today, but which also discusses the mathematical possibility of describing the mixed state of the original ensemble after preparation of the incident state preceding the measurement of $\underline{A}$, instead of in terms of the noncomplete set of states $\{\underline{a}'\}$, rather in terms of a complete orthonormal set $\{\tilde{a}\}$ diagonalizing the density matrix. This would replace (10b) by

$$\sum_{\tilde{a}} (\tilde{a},b)^2 = 1 \,, \tag{10c}$$

but we would keep

$$\sum_{\tilde{a}} \tilde{t}_{a\tilde{a}} \neq 1 \tag{12c}$$

similar to (12b). Moreover, the $\tilde{a}$ are not descriptive of the subensembles into which one would for physical reasons want to decompose the original ensemble.

TIME ARROW OF QUANTUM THEORY RELATED TO TIME ARROW OF OUR MIND

The moral of this story is that in the application of quantum theory there is the preferred time direction which is described mathematically by the difference between equalities (10a) or (12a) and inequalities (10b) or (12b-c). The inequalities (10b) are due to the fact that our gadgetry leads toward <u>selected</u> states <u>a'</u>, <u>b'</u>, ... which need not form complete orthogonal sets. On the other hand, the results of measurements are not known to us beforehand, so we do not select them individually, but hopefully use a mathematical scheme for determining a set of possibilities, for which the theory uses complete orthonormal sets {<u>a</u>}, {<u>b</u>}, and so on, guaranteeing equalities like (10a). The equalities (12a) express that there is conservation of probability as each selected incident state branches into a set of possible results for the following measurement. The inequalities (12b-c) are due to the fact that we can invent gadgetry that leads preferably to certain definite <u>a'</u> for an arbitrarily large set of states <u>a</u> entering the gadget.

Thus, the inequalities are brought about by the ingenuity with which we invent the gadgetry, while the equalities follow from rules which we assume to be valid for future results not beforehand known to us in detail, and our incapability or unwillingness to beat these rules by more inventiveness. Here, apparently, enters the time arrow of our mind. I will leave it to philosophers to formulate in a more precise way how this time arrow affects the problem at hand. As a physicist, I am satisfied by the fact that this influence is described mathematically by (10a-b) and (12a-c).

AFTERTHOUGHTS

<u>The paper presented in Detroit left some unanswered questions.</u>
<u>Here are some answers, some leading to new questions</u>:

AFTERTHOUGHT ONE.   MORE ABOUT RETARDED FIELDS.

Consider Fizeau's measurement of the velocity of light, using a lamp as a source of light, and using toothed wheels in the path of a light ray. The measurement proves the retarded nature of the light from the hot lamp toward the cool eye. That is, while Maxwell's equations allow advanced as well as retarded solutions for the light which has its source in the hot lamp, in fact the light arrives at the eye <u>after</u> it has been emitted by the lamp.

How could we have guessed without actually measuring the time delay, that, as time progresses, the photons would travel from the hot lamp toward the cool eye, rather than in opposite direction? If it had been true (as would be suggested by advanced fields) that our eye would see the lamp because it would be tickled by photons being sucked out of our eye by the lamp, then we would have a clear violation of the Second Law of thermodynamics by the heat (in the form of photons) traveling from our cool eye toward the hot lamp. If the Second Law derives its time arrow from the time arrow in our memory, it thus would follow that also the time arrow which in such thermodynamic (macroscopic) cases prefers retarded over advanced fields ultimately would derive from the time arrow of our mind.

Consider now a hot galaxy surrounded by space in which the radiation energy density corresponds to only a few degrees Kelvin. There is as little equilibrium between the galaxy and the surrounding space, as there was between the hot lamp and the cool eye. Again, we are in an improbable state that would become more probable if some heat were transported away from the hot galaxy toward its cooler surrounding. This time, however, the reason for the improbable state is not Fizeau's inventiveness in lighting a lamp, but it is, on the one hand, in the expansion of the universe diluting the radiation energy density in the surrounding space, and on the other hand in the heating up of the galaxy's stars by nuclear, gravitational, and possibly other processes. Due to the weakness of the electromagnetic coupling between this matter and the radiation field, the radiation that does take place has been insufficient to maintain an equilibrium.

Again, the Second Law of thermodynamics predicts that the galaxy will spit more photons into all directions, than it will suck out of its surrounding (which includes our faraway onlooking eye). As before, the Second Law claims that the radiative interaction between the galaxy and our eye or between the galaxy and some cool intergalactic dust should be described by retarded electromagnetic fields generated by the galaxy, rather than by advanced fields generated by it.

In this case, however, it is hard to believe that the time arrow in our faraway brains would be the reason why that galaxy is contracting while spitting out photons to heat up cold dust in space, rather than expanding at the cost of the energy of photons that it would be sucking out of the cool dust in its neighborhood. That is, how does the Second Law as it applies to this faraway galaxy know the direction of the time arrow of our mind?

Could it possibly be that the time arrow of our mind and of the life of other organisms is borrowed from a Mightier Time Arrow that puts its mark upon the entire universe we know?

AFTERTHOUGHT TWO. A CRITICISM OF OLBERS' PARADOX.

As the conditions for validity of Olbers' paradox are not at all satisfied according to relativistic cosmology, the effect of reversing expansion to contraction of the universe need not be at all the brightening of the sky to something brighter than the sun, as Olbers would have it. What the effect would be could be calculated for any cosmological model. We shall consider here a nonempty zero-pressure spherical universe that is at the present time t_n expanding, that at some time t_m will reach a maximum R_m of $R(t)$ of eq. (1), and that at some time $t*$ will have the same size $[R* \equiv R(t*) = R_n \equiv R(t_n)]$ as it has now, but then will be contracting. We want to compare the radiation density $U*$ then, as compared to its present value U_n. This density consists primarily of three parts: (1) its <u>primordial</u>

part, of at the present about 6×10^{-13} erg/cm^3 corresponding to a temperature of about $3°K$; (2) the part of it which has been emitted by stars of our own Milky Way, of which near the earth nearly all is due to the radiation emitted by the sun; and (3) the part that has been emitted, since the beginning of the universe at $t = 0$ with $R(0) = 0$, by the galaxies outside the Milky Way. As a function of the epoch $\underline{t}$, the first part depends merely on the instantaneous value of $R(t)$, and therefore has the same value of 6×10^{-13}erg/cm^3 at $t*$ as it has at the present t_n. The second part is not of a cosmological nature, and we assume it, too, to be the same at $t*$ as at t_n. As the radiative power of the sun is $L_\Theta = 3.8 \times 10^{33}$ erg/sec, near us at a distance $r = 1.5 \times 10^{13}$ cm from the sun there is an energy density

$$U_s = L_\Theta/4\pi r^2 c = 4.5 \times 10^{-5} \text{ erg/cm}^3$$

of solar radiation passing us at a velocity $\underline{c}$ directed away from the sun. Obviously, the primordial radiation with its much lower energy density is entirely negligible compared to this solar contribution.

The term which we question is the third one. For a spherical universe (with $k = +1$ in eq. (1)) we want a solution of the Friedmann equation which for zero cosmological constant is

$$(dR/dt)^2 = B/R(t) - c^2$$

with

$$B = (8\pi G/3) \rho R^3 = \text{constant}.$$

The observational data are crudely satisfied by a solution

$$R(t) = Q y(t)$$

with $y(t)$ given by

$$ct/Q = \text{arc cos } (1-y) - \sqrt{2y - y^2} \quad \text{for} \quad 0 < t < \pi Q/c,$$

$$ct/Q = 2\pi - \text{arc cos } (1-y) + \sqrt{2y - y^2} \quad \text{for} \quad \pi Q/c < t < 2\pi Q/c,$$

if we assume the present size of the universe to be about half the maximum size $2Q$ reached at $t_m = \pi Q/c$, and if we use

$$R_n = Q = 1.16 \times 10^{28} \text{ cm} = R*.$$

This gives

$$t_n = (\tfrac{1}{2}\pi - 1) Q/c = 6.98 \times 10^9 \text{ years}$$

for the present age of the universe, and gives

$$\rho_n = 2.4 \times 10^{-29} \text{ gram/cm}^3$$

for the effective total mass density responsible for the deceleration of this universe. (See the Ph.D. thesis of Shanker Raja, Purdue University, August 1970.) Then, $t* = (1 + 3\pi/2) Q/c$.

At any epoch t, let $N(t) = n/R(t)^3$ be the number of galaxies per unit volume, and let L be their average radiative power. For simplicity, assume n and L to be constants. For nL/R_n^3 we adopt here the value

$$N_n L \approx 10^{-31} \text{ erg/cm}^3\text{sec},$$

or possibly a bit more, but certainly not a hundred times more.

We merely want to establish an upper limit for the energy density of the radiation emitted by these galaxies since the big bang at $t = 0$. Therefore, we shall ignore here the fact that the galaxies have a finite size, so that faraway galaxies are partially covered up by closer ones, which might take away an appreciable fraction of the radiation reaching us, in particular in the calculation for the later time t^*. (If this covering up of the sky is taken into account, one finds the final result in the form of an integral that must be calculated numerically. Neglecting this effect, we obtain an integral that can be calculated analytically.) If this crude approximation would be applied to the theory of Olbers, we would find an infinite intensity of the sky, instead of the brightness of the sun. For our cosmological model we obtain a finite result, because we cannot look farther back than to the big bang.

We use, of course, the general-relativistic expression for the light reaching us, taking into account the effects of the metric eq. (1), including the redshift of spectra emitted before t_n and the blueshift observed at t^* of spectra emitted between t_n and t^*.

For the radiation densities at t_n and at t^* due to the emission by the galaxies we thus find

$$U_n = (\frac{3\pi}{4} - 2)\, \eta L/cQ^2 = 0.356\ N_n LQ/c = 0.14 \times 10^{-13}\ \text{erg/cm}^3,$$

$$U^* = (\frac{9\pi}{4} + 2)\, \eta L/cQ^2 = 9.068\ N_n LQ/c = 3.5 \times 10^{-13}\ \text{erg/cm}^3.$$

As mentioned, we may somewhat have underestimated the radiative power of the galaxies, but certainly not by a factor 100. We find that these densities are completely negligible compared to the radiation from the sun also at the time of contraction of the universe. There is no danger of burning up during this contraction, at least not in the model here adopted.

It is easily seen why our results differ so drastically from what Olbers predicted. In the first place, we must use the surface brightness not of the sun, but of a galaxy. The energy density corresponding to seeing the entire sky covered by stars of radiative power L and star radius R would be $L/\pi R^2 c$. For the sun this would be 8 erg/cm^3. For galaxies this may be 4×10^{-11} to 4×10^{-13} erg/cm^3. In the second place, not even the entire sky may be filled, as we can look back a finite time only. Finally, if expansion preceded contraction, light emitted far enough in the past would still be redshifted. (This would be the least important.)

The above calculations are, of course, do not reflect the dynamics of the universe, the creation of new galaxies, the disappearance of matter into black holes, or the emergence of new matter out of antiblackholes.

Throughout the preceding we have assumed that also during contraction of the universe we should use retarded fields, and not advanced fields that would make contraction look like expansion.

AFTERTHOUGHT THREE.   THE TIME ARROWS OF COSMOLOGY AND OF LIFE.

The question arises whether the time arrow of life is independent of the time arrow of cosmology. Though Olbers may be wrong and living beings may not burn up if they live in the direction of contraction of the universe, there may exist other reasons unknown to us which make life unstable in a contracting universe.

If this were so, of all life originating during contraction of the universe only that life would prosper which would regard this contraction as an expansion, observing redshifts of the spectra of distant galaxies by electromagnetic fields which it would consider to be retarded, though $\underline{we}$ would call them $\underline{advanced}$. At the epoch t_m of maximum size of the universe, the two kinds of life might meet, and kill off each other, if there were not quickly acting cosmological effects that would kill life in the wrong direction. Think of yourself meeting a man whose past lies in your future, and bashing his brain. How did he live with that bashed brain, in $\underline{his}$ time $\underline{before}$ you attacked him?

Maybe our direction of the time arrow of life can survive after the universe has started contracting. If opposite directions of the time arrows of life and of expansion of the universe are compatible, then perhaps $\underline{at}$ $\underline{the}$ $\underline{present}$ there exist $\underline{retro}$persons who receive their knowledge from our future. The question of what happens when we try to kill one of them is not the only paradoxical problem. When he and we compare the numbers of radial photon world lines around a hot star in cool space, each of us would want to claim that more lines go out than go in. (This is why each of us calls the fields retarded, while we mean opposite things by these words.) How can both of us be right? Is this question comparable to the question asked in quantum theory in connection with a measurement: Which state vector is correct, the one we use $\underline{before}$ the measurement, or the reduced state vector we use $\underline{after}$ it? That is, does the answer to the question lie in the $\underline{lack\ of\ reality}$ of fields we use in our humble human attempts to predict unpredictable nature?

$\underline{Or}$, could the answer possibly lie in the last sentence of Afterthought One? Do retropersons not exist, because $\underline{all}$ life was created after the image of the Big Time Arrow, called $\underline{Time}$ $\underline{Asymmetry}$ in the language of physics? It would save us from quarrels between persons and retropersons whether the coefficients 0.356 and 9.068 on the previous page, possibly should be interchanged at one of the two times t_n or t^* as the two meet and discuss the result they expect of a $\underline{measurement}$ of the instantaneous radiation density around them.

DISCUSSION

Editor's Note: The following was edited by both the editor and the speaker from tapes of the question and answer period which followed the talk.

172

NEWTON: I don't really understand why you say that there's something singled out in quantum mechanics about prediction rather than postdiction. It seems to me that the postulate is simply that the inner product says that if I know that the state is an eigenstate Ψa, then what is the probability for finding it in an eigenstate of some other variable b right now. From this I could now infer what it will be in the future if I know the development of the system in the future or I could equally well use the state as it has developed from the past and put in the way the state was in the past.

A. Well that's the way some people think--I show here that that leads to contradictions. And that the only way that you can do it is for the special cases where for the prediction you assume a future state that is garbled and for the postdiction you have a past state that is garbled. In such cases, I give examples of it, then it works, but I can also quote examples where people were led to completely nonsensical results just because they tried to do what you tried to do there, they blindly tried postdiction when the conditions for it were not satisfied; you just get nonsense.

NEWTON: That I would agree. Of course you have to know what you are doing. You have to know what this wavefunction describes, what kind of an ensemble it describes.

A. You have to have an ensemble which has a garbled initial state and then you're all right. But these people had a case where the initial state was not garbled. They had really picked the particular result of a measurement sometime in the past, and therefore it was no longer garbled, and then its faulty, and you could prove by it all kinds of things, because you have nonsense.

SCHLEGEL: I have two questions, first on your mathematical results. It seems to be a kind of quantum-mechanical H- theorem. You have the same result as an H-theorem, because you talk about information. Is this so?

A. I don't quite see the connection.

SCHLEGEL: Well the H-theorem tells us that there is a preferred direction in the evolution of a system.

A. No, the H-theorem tells us that if you go to a different time that your H goes down and therefore your probability goes up. It doesn't matter which direction you go.

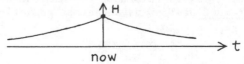

SCHLEGEL: Well, direction in time.

A. Yeah, but you can go forwards or backwards, it doesn't matter.

SCHLEGEL: The probability favors going forward.

A. No. That's what people put into it because they feel ashamed about the past. But that is not a proof. It's just an excuse, you see. You have to talk about Stosszahl Ansatz and things like that and you figure out some excuses and you give some examples of these excuses, but it is not a general logical structure there.

SCHLEGEL: Well, certainly statistically....

A. From a purely statistical point of view, the probability always likes to be large, or the H likes to be small, and therefore at any time the H is higher than it likes to be, whichever direction you go.

DRESDEN: It is precisely time inversion symmetric. What it depends on is how far you are away from an equilibrium-H. And if you are a certain distance away from that, then both in the future and also in the past you would be more likely to have been nearer by. You would have been more likely to have been at the higher point. But its precisely symmetrical in time.

SCHLEGEL: How do we then come to the result that we go from the H-theorem to a thermodynamic interpretation, that the entropy actually will increase?

DRESDEN: We do two things, but the main reason, the main way, the main trick that we pull at that point (but that is a different subject) is that we go to the thermodynamic limit. We make the system infinitely large. And we also then make a limit in terms of time. And in these two limits you have to be very careful how you handle it. You see how it is done; you really make the system infinitely large. Stated differently that makes a Poincaré cycle, effectively. You make a Poincaré cycle infinitely large and therefore you never get back away from it.

A. Yes. Such derivations are all swindles because these times are longer than the lifetime of the universe. Unless you have a cyclic system, I would simply say that there is no logical derivation and you just hope for the best because you like to have the Second Law so why wouldn't I just assume it. You use it towards the future. No one is going to use it towards the past. No harm is done by making that statement about the past.

SCHLEGEL: Now this relates to the experience. The other question that I wanted to ask: One of your first comments was that memory enables us to state the Second Law. Is that something of _that_ sort?

A. I say that we use the lack of knowledge of the future as the
reason why we <u>assume</u> (we don't really ever prove it) that the most
likely state is going to be realized. We say its OK, I can't imagine that an unlikely state would occur. And in class one always
gives numerical examples and they are by fantastic factors 1010^{24}
or something, more likely. Who is going to assume that that unlikely state will occur. You'd say I must be crazy to assume that.
And that is the only reason for the Second Law.

SCHLEGEL: Then, you are using memory as a way of defining a
direction?

A. Yeah. Say it is because we do not know the future. The past
you might have known, you should have known, you should have remem-
bered. You should have known there was a very specific state there,
and which therefore has a very small likelihood, because it was a
very particular state. And its only if you don't know the state,
you have a big probability with many microstates in your ensemble.

SCHLEGEL: Yes, there are those who would say that the Second Law
has a more substantial role however, that in fact the Second Law
gives us a way of describing what the direction of the natural process would be, such that one wouldn't put the memory as a prior
bound.

A. That is the way you use it in practice. But I say where do you
get this from? You get this from the fact that you don't know the
future. So, I am saying that basically this is related to the
direction of our memory.

SCHLEGEL: And therefore, if we in fact found that entropy usually
decreased, you would say--well we can have it stated that way too,
because the memory would still give a unique direction in time.

A. It will decrease if your ensembles which describe the state
would be far more precise in the future than they are now. It
seems rather unlikely that you would know more precisely what
happens in the future than what you know now. But the person
who lives backwards in time, for him that would be true, possibly.

ROLNICK: In your initial comments you used the word ensemble for a
statistical mechanics kind of system, temperature, and then you
used the word ensemble when you were dealing with wavefunctions.
It seems to me that they're intrinsically different. One ensemble
is a measure of ignorance and the other one is the necessary
ensemble which is required.

A. There is a little bit more variety in the quantum mechanical
ensemble than in the classical one. I agree with that. There are
some special properties that classically you don't assign and quantum mechanically you have, but I would still call it an ensemble.
I would say this classical ensemble is a special case, perhaps, of
the interference terms. For the density matrix is diagonalized or
something like that, you see. Right after measurement if we reduce a wavefunction, then we are in a case where we assume that for
all practical purposes we may neglect the off-diagonal elements
and then the two coincide, which is of course an approximation,
but is a good one.

176

LIST OF PARTICIPANTS

Adel Antippa - Université du Québec, Trois Rivières, Québec, Canada
Debojit Barua - Wayne State University, Detroit, Michigan
F.J. Belinfante - Purdue University, Lafayette, Indiana
John Bolzan - Ohio State University
Vimal Chowdhry - Wayne State University, Detroit, Michigan
Thomas Cobb - Bowling Green State University, Bowling Green, Ohio
James H. Cooke - University of Texas, Arlington, Texas
Jerry Crandell - Bowling Green State University, Bowling Green, Ohio
James H. Crichton - Seattle Pacific College, Seattle, Washington
Robert Deck - University of Toledo, Toledo, Ohio
Harry Denman - Wayne State University, Detroit, Michigan
William Dorenbusch - Wayne State University, Detroit, Michigan
Max Dresden - SUNY, Stony Brook, New York
Kenneth Duff - Wayne State University, Detroit, Michigan
George Duncan - Bowling Green State University, Bowling Green, Ohio
Allen Everett - Tufts University, Medford, Massachusetts
William Fadner - University of Northern Colorado, Greeley, Colorado
L. D. Favro - Wayne State University, Detroit, Michigan
David Fradkin - Wayne State University, Detroit, Michigan
Robert Goble - University of Utah, Salt Lake City, Utah
Peter Havas - Temple University, Philadelphia, Pennsylvania
Ernest Henninger - De Pauw University, Greencastle, Indiana
Ira Herbst - University of Michigan, Ann Arbor, Michigan
Ellis Kane - Detroit, Michigan
Patrick Kenealy - Wayne State University, Detroit, Michigan
Peter Kirschenmann - Wayne State University, Detroit, Michigan
Marylou Larsen - Michigan State University, East Lansing, Michigan
Amos Lev - Wayne State University, Detroit, Michigan
Charles Lincoln - SUNY, College of Fredonia, Fredonia, New York
D.P. Majumdar - University of Michigan, Ann Arbor, Michigan
Louis Marchildon - Université du Québec, Trois Riviére PQ, Canada
Frank Meyer - Wisconsin State University, Superior, Wisconsin
Charles Miller - Trinity College, Hartford, Connecticut
Cyrus Moazed - Wayne State University, Detroit, Michigan
Lowell Morgan - Wayne State University, Detroit, Michigan
Kenneth Mucker - Bowling Green State University, Bowling Green, Ohio
Roger Newton - Indiana University, Bloomington, Indiana
Kenneth Nordtvedt - Montana State University, Bozeman, Montana
Karl Parsons - Eastern Michigan University, Ypsilanti, Michigan
Roger Ptak - Bowling Green State University, Bowling Green, Ohio
William Rolnick - Wayne State University, Detroit, Michigan
Mark A. Runkle - Powell, Ohio (Ohio State University)
Alvin Saperstein - Wayne State University, Detroit, Michigan
Richard Schlegel - Michigan State University, East Lansing, Michigan
Joseph Shpiz - CUNY, City College, New York
Scott Smith - East Illinois University, Charleston, Illinois
Richard Spector - Wayne State University, Detroit, Michigan
Henry Stapp - Lawrence Berkeley Laboratory, University of California,
    Berkeley, California

Talbert Stein - Wayne State University, Detroit, Michigan
Ronald Stoner - Bowling Green State University, Bowling Green, Ohio
E.C.G. Sudarshan - University of Texas, Austin, Texas
Paul Tipler - Oakland University, Rochester, Michigan
Paul Weiss - Wayne State University, Detroit, Michigan
David Williams - University of Michigan, Ann Arbor, Michigan

AIP Conference Proceedings

		L.C. Number	ISBN
No. 1	Feedback and Dynamic Control of Plasmas (Princeton 1970)	70-141596	0-88318-100-2
No. 2	Particles and Fields - 1971 (Rochester)	71-184662	0-88318-101-0
No. 3	Thermal Expansion - 1971 (Corning)	72-76970	0-88318-102-9
No. 4	Superconductivity in d- and f-Band Metals (Rochester 1971)	74-188879	0-88318-103-7
No. 5	Magnetism and Magnetic Materials - 1971 (2 parts) (Chicago)	59-2468	0-88318-104-5
No. 6	Particle Physics (Irvine 1971)	72-81239	0-88318-105-3
No. 7	Exploring the History of Nuclear Physics (Brookline 1967, 1969)	72-81883	0-88318-106-1
No. 8	Experimental Meson Spectroscopy - 1972 (Philadelphia)	72-88226	0-88318-107-X
No. 9	Cyclotrons - 1972 (Vancouver)	72-92798	0-88318-108-8
No.10	Magnetism and Magnetic Materials - 1972 (Denver)	72-623469	0-88318-109-6
No.11	Transport Phenomena - 1973 (Brown University Conference)	73-80682	0-88318-110-X
No.12	Experiments on High Energy Particle Collisions - 1973 (Vanderbilt Conference)	73-81705	0-88318-111-8
No.13	π-π Scattering - 1973 (Tallahassee Conference)	73-81704	0-88318-112-6
No.14	Particles and Fields - 1973 (APS/DPF Berkeley)	73-91923	0-88318-113-4
No.15	High Energy Collisions - 1973 (Stony Brook)	73-92324	0-88318-114-2
No.16	Causality and Physical Theories (Wayne State University, 1973)	73-93420	0-88318-115-0
No.17	Thermal Expansion - 1973 (Lake of the Ozarks)		0-88318-116-9
No.18	Magnetism and Magnetic Materials - 1973 (Boston)		0-88318-117-7
No.19	Physics and the Energy Problem - 1974 (APS Chicago)		0-88318-118-5